Andreas Eisele

Millimeter-Precision Laser Rangefinder Using a Low-Cost Photon Counter

Karlsruhe Series in Photonics & Communications, Vol. 16
Edited by Profs. J. Leuthold, W. Freude and C. Koos

Karlsruhe Institute of Technology (KIT)
Institute of Photonics and Quantum Electronics (IPQ)
Germany

Millimeter-Precision Laser Rangefinder Using a Low-Cost Photon Counter

by
Andreas Eisele

Dissertation, Karlsruher Institut für Technologie (KIT)
Fakultät für Elektrotechnik und Informationstechnik, 2013

Impressum

 Scientific Publishing

Karlsruher Institut für Technologie (KIT)
KIT Scientific Publishing
Straße am Forum 2
D-76131 Karlsruhe

KIT Scientific Publishing is a registered trademark of Karlsruhe
Institute of Technology. Reprint using the book cover is not allowed.

www.ksp.kit.edu

Print on Demand 2014

ISSN 1865-1100
ISBN 978-3-7315-0152-7

Millimeter-Precision Laser Rangefinder Using a Low-Cost Photon Counter

Zur Erlangung des akademischen Grades eines

DOKTOR-INGENIEURS

von der Fakultät für
Elektrotechnik und Informationstechnik
des Karlsruher Instituts für Technologie (KIT)

genehmigte

DISSERTATION

von

Andreas Eisele, M. Sc.

geboren in Stuttgart

Tag der mündlichen Prüfung: 06. Dezember 2013
Hauptreferent: Prof. Dr. Dr. h.c. Wolfgang Freude
Korreferent: Prof. Dr. Marc Weber

Fachlicher Betreuer bei BOSCH   Dr. Bernd Schmidtke

To my family

Wahrlich, es ist nicht das Wissen, sondern das Lernen,
nicht das Besitzen, sondern das Erwerben,
nicht das Dasein, sondern das Hinkommen,
was den größten Genuss gewährt.

Carl Friedrich Gauß, Schreiben an Wolfgang Bolyai
Göttingen, 2. September 1808

Abstract

This work relates to the field of distance measurement, more precisely to optical distance measurement with a laser rangefinder. Laser rangefinders are electro-optical measurement devices for measuring the distance to an object. The non-military target application in this work is measuring distances on a construction site. The target distance ranges from few centimeters to tens of meters. Millimeter distance accuracy is required. A further constraint that has to be met is a low overall system cost. Construction or DIY (do it yourself) applications are a very price sensitive market; hence the overall system cost should be as low as possible.

We successfully demonstrate a millimeter-precision laser rangefinder using a low-cost photon counter. The measurement is based on the time of flight of a laser beam to the target and back to the rangefinder. A distance accuracy of 1 mm requires that the time of flight is determined with 6.7 ps timing accuracy. The heart of this work is an application-specific integrated circuit (ASIC). The ASIC comprises a modulator for the laser current, photodetectors and timing circuitry. Single-photon avalanche diodes (SPADs) are used as the photodetectors. Conventional laser rangefinders use avalanche photodiodes (APDs). The output of a SPAD is an electrical pulse for each detected photon. Since SPADs have recently been presented in a CMOS-compatible imaging technology, the ASIC can be manufactured at large scale and at low-cost in existing semiconductor fabrication lines. For the timing circuitry, a novel binning architecture for sampling the received signal is proposed in this work. This binning architecture mitigates non-idealities which are inherent to a system with SPADs and timing circuitry in one chip. A theoretical system performance simulation estimates the performance for a SPAD-based laser rangefinder in various scenarios. Laboratory measurements and a successful experimental demonstration of a prototype laser rangefinder conclude this work. The prototype reaches the performance target, namely distance measurement with millimeter precision.

In conclusion, a SPAD-based laser rangefinder has the potential to outperform a conventional APD-based system in terms of cost, size and manufacturability while reaching a performance level that suits most application scenarios on a construction site.

Achievements

In this work, we study a laser rangefinder using a low-cost photon counter. The goal is measuring distances with millimeter precision over a target distance ranging from a few centimeters to tens of meters. Single-photon avalanche diodes (SPADs) or at least the basic underlying effects are well-known since the 1960s. However, only recently first devices have been demonstrated in a CMOS technology, which known to be suitable for large-scale manufacturing of cell-phone cameras. Because this type of photodetector is CMOS-compatible, it can be co-integrated with further circuitry, which promises a low overall system cost. In the following, the main achievements of this work are summarized.

Single-photon Avalanche Diodes

- **Characterization of SPAD Parameters:** We study the performance parameters of single-photon avalanche diodes (SPADs), in particular the photon detection efficiency (PDE), the dark count rate (DCR), the dead time, afterpulsing and time jitter. Various device variations are provided by our technology partner STMicroelectronics in a 130 nm imaging technology. Several parameters are characterized over an array of 3×3 SPADs in a first wafer run and over an array of 6×10 SPADs in a second wafer run.

- **SPAD Saturation and Drive Circuit Options:** Because of their capability to detect single photons, SPADs are typically employed in low-light laboratory applications such as fluorescence lifetime microscopy. However, in the target application of a laser rangefinder, the detector has to work under ambient light conditions. Instead of detecting count rates of few photons per second, we operate SPADs at high count rates in this research and drive the devices into saturation. The saturation behavior is analyzed with different types of quenching circuits that reset the device after an event (incident photon) has been detected. Passive and active quenching circuits are characterized in this work. To the best of our knowledge, the 185 MHz count rate reached with an active quenching circuit has set a new benchmark for maximum count rate.

  Furthermore, we operate SPADs using positive and negative drive circuit options. With the positive-drive option, the capacitance of the SPAD floating node reduces dramatically, such that the maximum count rate in passive quenching is increased from 11.9 MHz with negative-drive option to 45.6 MHz with positive-drive option.

- **SPAD Parameters for Laser Rangefinder:** In the course of this work, we identify several SPAD parameters of particular relevance for a ranging system. While the dark count rate and afterpulsing should be as low as possible for certain laboratory measurements, these parameters are less important for a laser rangefinder. In a SPAD-based laser rangefinder, however, any influence that affects SPAD timing, i.e., the time between photon incidence and an electrical pulse at the output, is crucial. Moreover, a high stability of the temporal behavior versus device mismatch, temperature, and voltage is of utmost importance.

System Aspects of a Laser Rangefinder

- **Performance Estimation:** We model the performance of a single-spot, SPAD-based laser rangefinder with a fixed-focus optical system. An optical analysis determines the expected light levels on the detector plane, including background light and signal light that is scattered back from the target. Experimentally determined SPAD parameters from the previous section are taken into account. Distance measurement is based on frequency-domain reflectometry, which is also referred to as single-photon synchronous detection (SPSD) when using SPAD detectors. To achieve millimeter-precision distance measurements, a modulation frequency of about 400 MHz to 1.1 GHz is required. A Monte Carlo simulation estimates non-idealities of the so-called "binning architecture" (an architecture for sampling windows in time) for evaluating the received backscattered signal.

- **Bin Homogenization:** For sampling and evaluating the signal that is backscattered form the target, we propose a so-called "bin homogenization" scheme to mitigate non-idealities of the sampling windows of the binning architecture without prior calibration. The semiconductor process used for the fabrication of SPADs is optimized for optical properties, however, it suffers from severe RF-limitations. Nevertheless, an RF modulation frequency is required to achieve millimeter precision. A high-speed oscillator to drive a clocked shift-register to generate the sampling windows for sampling the incident signal is not available in this technology. Instead, we derive the individual sampling windows (bins) from the delay stages of a voltage-controlled oscillator. Because of non-ideal matching of the delay stages, the derived bins (sampling windows) have unequal bin widths that must be compensated for. State-of-the-art measurement schemes would require calibration for each device before each new measurement series. We overcome this limitation with the proposed bin homogenization scheme (patent-pending).

- **Laser Rangefinder ASIC:** We combine SPADs, timing-circuitry and a laser modulator into a system-on-chip laser rangefinder ASIC. Our first multi-project wafer (MPW2009) comprises the first system-on-chip laser rangefinder including a 3×3 SPAD array, timing-circuitry and a laser driver on a single ASIC. Our second multi-project wafer (MPW2011) comprises a 6×10 SPAD array and several improvements to building blocks. Most importantly, it successfully demonstrates the proposed bin homogenization scheme.

  Both chips are designed and implemented by our technology partners Bosch Automotive Electronics and STMicroelectronics.

- **Laser Rangefinder Prototype:** We successfully build a laser rangefinder prototype based on the laser rangefinder ASIC including SPADs. Under ideal laboratory conditions and with the proposed bin homogenization scheme, the laser rangefinder prototype reaches a maximum distance error as low as 0.24 mm for a target distance of up to 3.2 m, even with unavoidable noise and systematic errors. To the best of our knowledge, we achieved the highest accuracy for a SPAD-based ranging system so far.

Zusammenfassung (German Abstract)

Die vorliegende Arbeit befasst sich mit einem Laser-Messgerät zur Entfernungsmessung. Aus militärischen Anwendungen sind Laser-Entfernungsmessgeräte insbesondere zur Zielerfassung bekannt. Die vorliegende Arbeit befasst sich jedoch mit der zivilen Nutzung handlicher Laser-Entfernungsmessgeräte für Heimwerker, Handwerker und zum Einsatz auf Baustellen. Typische Zielentfernungen liegt im Bereich von wenigen Zentimetern bis zu einigen zehn Metern. Hierbei ist insbesondere bei kurzen Entfernungen eine Messgenauigkeit im Millimeterbereich erforderlich. Eine für den wirtschaftlichen Erfolg entscheidende Randbedingung sind geringe Kosten, da preissensible Märkte bedient werden sollen.

Die vorliegende Arbeit demonstriert erfolgreich ein Laser-Entfernungsmessgerät mit einer Messgenauigkeit im Millimeterbereich. Das Gerät arbeitet nach dem Lichtlaufzeitverfahren, wobei die Laufzeit eines Laserstrahls zum Ziel und zurück zum Entfernungsmessgerät ausgewertet wird. Dabei erfordert eine Messgenauigkeit von 1 mm eine Zeitauflösung von 6.7 ps. Als Lichtsensor findet ein kostengünstiger Photonenzähler Verwendung.

Das Kernstück des Entfernungsmessgeräts ist eine anwendungsspezifische integrierte Schaltung (ASIC, application specific integrated circuit). Der ASIC vereint sowohl die Zeitmess-Elektronik als auch die Photodetektoren auf einem einzigen Chip. Als Photodetektoren kommen integrierte Einzelphotonen-Lawinendetektoren (SPAD, single-photon avalanche diode) zum Einsatz, wohingegen konventionelle Laserentfernungsmessgeräte diskrete Lawinen-photodetektoren (APD, avalanche photodiode) verwenden. Das Ausgangssignal einer SPAD ist ein elektrischer Puls für jedes detektierte Photon. Diese Pulse können ohne zusätzliche Verstärker oder Filter unmittelbar weiterverarbeitet werden. Da SPADs kürzlich in einer CMOS-kompatiblen Halbleiter-Fertigungstechnologie implementiert wurden, kann der ASIC in Großserie mit bereits bestehenden Halbleiter-Fertigungsanlagen zu geringen Stückkosten produziert werden.

Für die Zeitmess-Elektronik wird im Rahmen dieser Arbeit eine neuartige 'Binning'-Architektur zum Abtasten des empfangenen, vom Zielobjekt rück-reflektierten Signals vorgeschlagen. Diese Binning-Architektur entschärft

Probleme, welche bei hochaufgelöster Abtastung einem integrierten System mit SPADs und Zeitmess-Elektronik auf einem Chip innewohnen. Eine theoretische Performance-Abschätzung simuliert die Leistungsfähigkeit des vorgeschlagenen SPAD-basierten Laser-Entfernungsmessgeräts bei verschiedenen Messbedingungen. Versuche im Labor sowie erfolgreiche Tests mit einem Prototyp eines Laser-Entfernungsmessgeräts runden diese Arbeit ab. Der Prototyp erreicht die Zielvorgabe: Entfernungsmessung mit Millimeter-Genauigkeit.

Das demonstrierte SPAD-basierte Laser-Entfernungsmessgerät übertrifft ein konventionelles APD-basiertes System in Bezug auf Kostengünstigkeit, Handlichkeit der Abmessungen und Fertigungsfreundlichkeit. Die Messgenauigkeit erfüllt die Anforderungen der relevanten Mess-Szenarien für Handwerk und Heimwerk.

Contents

1 Introduction

The present work relates to the field of distance measurement, more precisely to optical distance measurement with a laser rangefinder. This chapter provides the motivation, gives the measurement task and measurement condition and concludes with an overview of the organization of this thesis.

1.1 Motivation

> "Measurement has been important ever since man settled from his nomadic lifestyle and started using building materials; occupying land and trading with his neighbors. As society has become more technologically orientated much higher accuracies of measurement are required in an increasingly diverse set of fields, from micro-electronics to interplanetary ranging." [1]

It is not exaggerated to say that measured lengths or distances are omnipresent in our everyday lives. The fitting of clothes, the matching sizes of nuts and bolts, the height of doors and steps as well as distances between different places are just a few examples where we rely on accurate length or distance measurements. Distances are measured by direct or indirect comparison with a reference. One of the oldest techniques for distance measurement is using body parts as a reference. The 'cubit', defined as the length of the arm from the tip of the finger to the elbow, has been used by the ancient Egypts in 3,000 BC [1]. Since such a measure varies considerably for different people a standardized distance unit must be defined. Today's definition of a meter is about 30 years old [2]. In the 17th General Conference on Weights and Measures in 1983, the meter has been defined as the length of the path travelled by light in vacuum during a time interval of 1/299,792,458 of a second. Thus, the meter is defined by the speed of light and a *time of flight*.

The two main criteria for a distance measurement are *range* and *accuracy*. The range defines the minimum to the maximum distance that can be measured. The accuracy defines how reliable a distance can be measured and how fine the resolution is. It should be noted that in many applications only the relative accuracy with respect to the target distance is important.

The appropriate technical means for distance measurement depend on the application. Examples for distance measurement techniques [3] include (from small to large range, wherein some ranges overlap): optical interferometry, depth of focus of an optical system, direct measurement by comparison with a reference, triangulation, time-of-flight methods / LIDAR (LIght Detection And Ranging), RADAR (RAdio Detection And Ranging) and evaluation of redshift at interstellar distances. A more detailed overview of optical distance measurement techniques is provided at the beginning of Chapter 2.

Besides the main criteria, range and accuracy, further technical and non-technical requirements may have to be considered. For example if a contact (e.g. using a caliper or yardstick for direct reference) or contactless (interferometry, depth of focus, triangulation, radar) measurement is possible or desired.

Furthermore, economic aspects are of utmost importance for a successful product. The usability and acceptance of a product on the market also depend on the design, including size and weight as well as the user interface. Furthermore, one of the most important factors is the cost of the distance measurement system.

One particularly price-sensitive market is the construction industry. Even though laser rangefinders are commercially available for military applications and in construction, their market penetration in the construction industry, by craftsmen and in the DIY (do it yourself) sector is rather limited. When it comes to measuring distances on a construction site the most common tool today is still a tape measure or a carpenter's ruler/yardstick. A low system cost is crucial to open up new, price-sensitive market segments.

Regarding low-cost rangefinders, the low-price segment is dominated by ultrasonic rangefinders. These sonar systems evaluate the time of flight of an acoustic wave to a target and back. The speed of sound in air of 340 m/s is much lower than the speed of light of almost 300,000 km/s which is used in a laser rangefinder. Therefore, a sonar system can use cheap low-frequency electronics such as standard microcontrollers with µs timing capability to achieve millimeter precision. However, ultrasonic rangefinders fail to provide reliable distance measurements in measurement scenarios with obstacles as shown in Fig. 1–1 (a).

For a reliable distance measurement the starting point and the ending point must be known precisely. In sonar rangefinders, an additional non-modulated laser beam, installed as an aiming aid, creates the illusion of measuring the distance to a well-defined spot. However, any obstacle in the acoustic signal cone (Fig. 1–1 (a)) reflecting a dominant echo might be measured. In a true laser rangefinder (Fig. 1–1 (b)), the laser beam constitutes both aiming aid and measuring medium. In a multitude of measurement scenarios, like Fig. 1–1 (a) and (b), this is the only approach to produce reliable distance measurement results. Unfortunately, today's laser rangefinders are significantly more expensive than ultrasonic rangefinders.

Nonetheless, there are also limitations on the application scenarios of laser rangefinders. The distance to a transparent target, such as a window, cannot be measured since the laser beam passes through (see Fig. 1–1 (c)). A reflective target can deflect the laser beam such that the distance to a different object is measured as shown in Fig. 1–1 (d).

Recent advances in semiconductor technology provide a promising basis for miniaturization and cost-reduction of true laser rangefinders. The target of this dissertation is reliable distance measurement with a true laser rangefinder at low cost. In a first step, we evaluate single-photon avalanche diodes (SPADs), a type of CMOS compatible low-cost photon counter, as a prospective type of detector for a low-cost ranging system. In a second step, we evaluate the feasibility of a millimeter-precision laser rangefinder using SPADs.

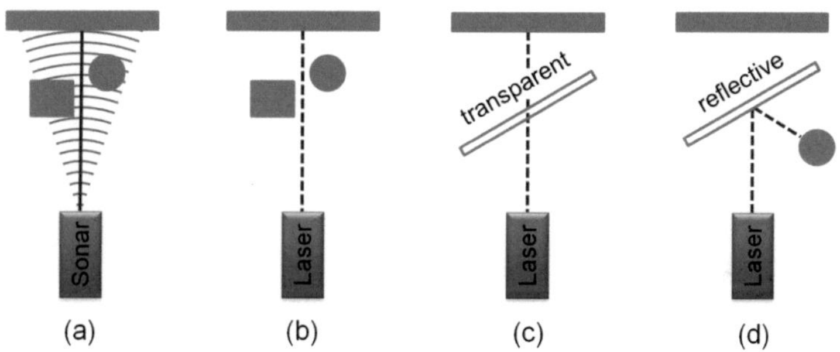

Fig. 1–1: Measurement scenarios (a) with obstacles using a sonar rangefinder with laser aiming aid, (b) with obstacles using a laser rangefinder, (c) laser rangefinder on transparent target, (d) laser rangefinder on reflective target.

1.2 Measurement Task and Measurement Conditions

The target application for a laser rangefinder in this thesis is distance measurement on a construction site. Fig. 1–2 illustrates exemplary measurement scenarios.

In Fig. 1–2 (a) the width of a hallway is measured. This task could also be performed with a conventional yardstick (typically 2 m long) and does not necessarily require a laser rangefinder. However, for measuring distances that are considerably longer than the yardstick, such as the length of a hallway, a distance measurement with a yardstick is not efficient. The use of a rangefinder saves time and money in such a scenario.

In Fig. 1–2 (b) a diagonal of a staircase is measured. In this scenario a direct measurement with a yardstick is impossible since the measurement is performed over an opening in midair.

In Fig. 1–2 (c) the length of a roof element (solid red line) is determined. Since the roof element is out of reach, the length of the element cannot be measured directly. Instead, the lengths of the dashed red lines are measured and stored in a memory of the laser rangefinder. The length of the roof element is calculated therefrom by the laser rangefinder. A craftsman saves time since all operations are performed by the laser rangefinder.

Furthermore, in the high-price segment combined distance / angle measuring and subsequent trigonometric calculations are state of the art [4, 5] for measuring distances without direct line of sight between the laser rangefinder and the target.

The required measurement range on a construction site is between some centimeters to tens of meters. The required accuracy again depends on the specific task. However, over a distance of some meters, micrometric accuracy is certainly not required on a construction site since, for example, the thickness of plaster or the positioning of a wall have an uncertainty of several millimeters anyway. Instead of ultimate precision, there is rather a demand for lightweight, handheld devices with millimeter accuracy.

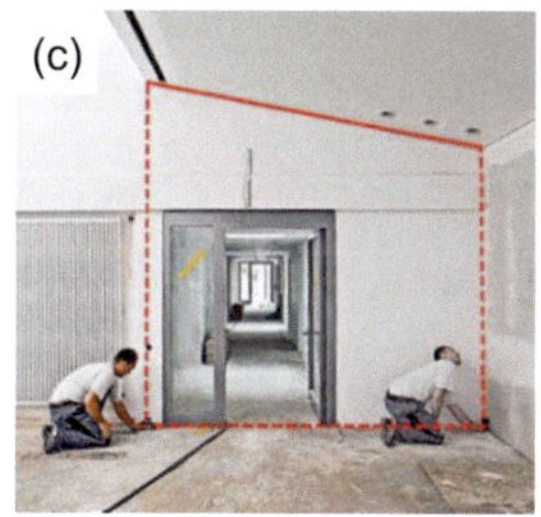

Fig. 1–2: Measurement scenarios. (a) width of hallway, (b) diagonal of staircase (solid red line), (c) length of roof element (photos by courtesy of Robert Bosch GmbH)

Fig. 1–3 depicts examples of commercially available laser rangefinders. Fig. 1–3 shows a high-end laser rangefinder with a measurement range up to 250 m (Bosch GLM 250 VF Professional [4]) and a compact second laser rangefinder with a measurement range up to 80 m (Bosch GLM 80 Professional [5]).

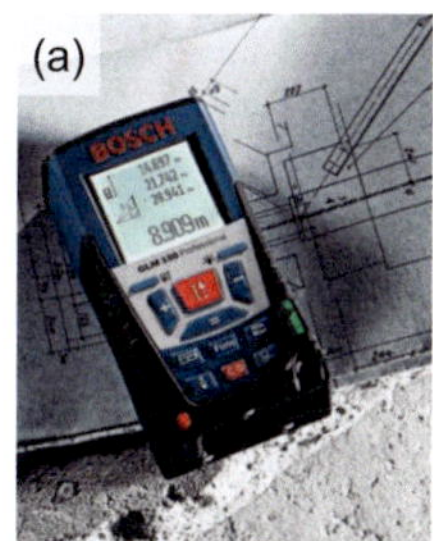

Fig. 1–3: Examples of commercially available laser rangefinders (a) Bosch GLM 250VF (b) Bosch GLM 80 (photos by courtesy of Robert Bosch GmbH)

The typical performance metrics are listed in Table 1.1, second column, for a Bosch GLM 250 [4]. This device targets the professional market. The maximum measuring range of up to 250 m enables distance measurements on large construction sites such as conference buildings, train stations or football stadiums. The high background light tolerance of this device also enables outdoor measurements. In many application scenarios, however, the requirements are much lower. For most indoor and DIY (do it yourself) applications a range L up to 10...20 m, a relaxed accuracy and a limited background light tolerance of 1,000-10,000 lx will be sufficient. A light level of 100,000 lx corresponds to direct sunlight, a light level of 1,000 lx corresponds to

TV-studio lighting [6]. The requirements for the target system in this research are specified in the rightmost column of Table 1.1.

The measurement accuracy of a laser rangefinder depends on the true target distance L. For example, noise contributions will increase at a large target distance due to a lower signal-to-noise ratio. The specified range dependent accuracy $\xi(L)$ lies within the boundaries given by

$$\xi(L) = \pm(\xi_0 + \xi_1 L).$$ (1.1)

This definition comprises a range independent offset ξ_0, e.g. 1.5 mm, and a slope error ξ_1, e.g. 0.05 mm/m (see Table 1.1, second column). The specified boundaries at a true target distance L cover at least the systematic distance error μ_L and twice the non-systematic distance error σ_L. Further details about the measurement accuracy and a definition of systematic and non-systematic distance error can be found in Appendix A.

	Bosch GLM 250 [4]	**Target system**
Range L	0.05...250 m	0.05...20 m
Range dependent accuracy $\xi(L)$	$\pm\left(1.5\text{mm} + 0.05\,\dfrac{\text{mm}}{\text{m}} \times L\right)$	$\pm\left(2.0\text{mm} + 0.1\,\dfrac{\text{mm}}{\text{m}} \times L\right)$
Measurement time	typ. < 0.5 s, max. 4 s	same as GLM 250
Background light tolerance	100,000 lx	1,000-10,000 lx
Laser	1 mW, Class II, eye safe	1 mW, Class II, eye safe
Housing	dust and splash water protected, shock resistant according to the IP54 protection standard	n.a.

Table 1.1: Laser rangefinder performance data

Eventually, a laser rangefinder will be deployed in the harsh environment of a construction site. The IP54 standard (see IEC 60529) ensures dust and splash water protection. Even after mechanical shock the optical alignment of the unit has to remain unchanged. Test procedures include a drop test on a concrete floor which has to be expected on a construction site. Today's laser rangefinders are manufactured under cleanroom conditions and the optical system is aligned with μm-precision for each individual device. Besides reducing the cost of the system components by employing an application-specific integrated circuit (ASIC) with integrated single-photon avalanche diodes (SPADs), there is an enormous potential to reduce manufacturing cost, if the alignment could be simplified.

Eye safety, as a mandatory constraint, limits the average laser output power of the laser rangefinder to 1 mW [7].

1.3 Organization of Dissertation

The aim of this research is the design and development of a low-cost millimeter-precision laser rangefinder. This present work is a feasibility study for a laser rangefinder using single-photon avalanche diodes (SPADs).

Chapter 2 gives an introduction to different optical distance measurement techniques. Time-domain and frequency-domain reflectometry, in the context of ranging systems also referred to as direct and indirect time-of-flight distance measurement, are explained. The current APD-based system (APD = avalanche photodiode) is described as well as its limitations with respect to cost, size and manufacturing. Several integrated ranging systems and their performance data are presented. Single-photon synchronous detection (SPSD) has been identified as one possible solution to overcome the limitations of state-of-the-art laser rangefinders.

Chapter 3 focuses on CMOS single-photon avalanche diodes (SPADs) as an enabling technology that allows co-integration of the photodetector with laser driver, timing- and control-circuitry in a single ASIC. The basic device parameters of SPADs are described along with the necessary electronic circuitry. When pushing for millimeter accuracy, timing influences of few picoseconds must be considered.

Chapter 4 presents an experimental characterization of SPAD test structures.

Chapter 5 lays out the system concept of a millimeter-precision low-cost laser rangefinder based on single-photon synchronous detection (SPSD). The target of this chapter is an estimation of the achievable system performance. The optical power budget and the expected photon rates are estimated. A so-called 'binning architecture' will be introduced that is used for sampling the received signal. Non-idealities of the binning architecture at higher frequencies are considered along with timing drifts and noise. A novel patent-pending 'bin homogenization scheme' mitigates the problem of unequal bin widths. The system performance is estimated with a combination of analytical calculations and a Monte Carlo simulation.

<u>Chapter 6</u> presents the results of a system-on-chip laser rangefinder with SPADs, integrated laser driver and readout circuitry. The chip is implemented in a standard 130 nm CMOS imaging process. Two multi-project wafer (MPW) runs in 2009 and 2011 verified the concepts outlined in Chapter 5. A laboratory setup of a laser rangefinder has successfully been realized. Under laboratory conditions with weak environmental light, the distance error with bin homogenization is as low as 0.24 mm including noise and systematic errors.

Conclusions are provided at the end of Chapters 2, 4, 5 and 6.

2 Distance Measurement

This chapter provides an introduction to optical distance measurement. Different optical distance measurement techniques are classified in Section 2.1. Reasons are given why reflectometry is the most appropriate technique for the application of distance measurement on a construction site with a compact, handheld and low-cost device. Reflectometry is described in more detail in Section 2.2. Examples for state-of-the art ranging systems based on reflectometry are provided in Section 2.3.

2.1 Classification of Optical Distance Measurement Techniques

This section gives an overview of different active and passive optical distance measurement techniques. They are assessed with respect to single spot measurement scenarios on a construction site. Optical distance measurement techniques can be grouped into passive and active systems as shown in Fig. 2–1. A more detailed classification of different ranging techniques including scanner based methods for 3D image acquisition can be found in [8] and [9]. Passive systems only evaluate light coming from the target whereas active systems also feature a light source. The techniques are discussed below.

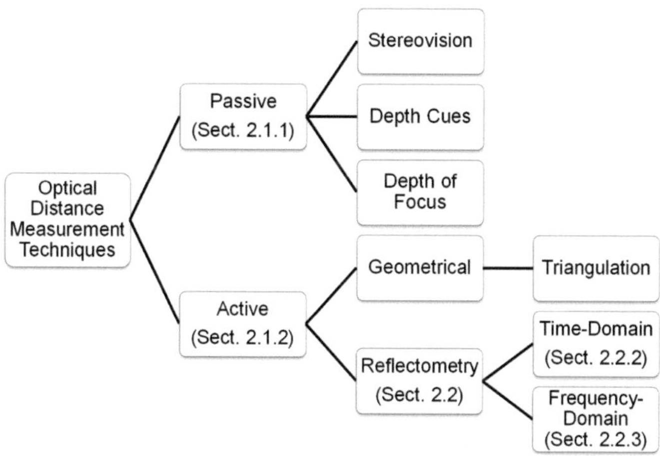

Fig. 2–1: Classification of optical distance measurement techniques. Further details are provided in the sections indicated in parentheses.

2.1.1 Passive Techniques

The fact that passive optical distance measurement systems do not require a light source is beneficial for a low-cost system. Standard commercially available image sensors can be used.

In <u>stereovision</u>, the disparity between two images taken with known camera spacing is evaluated. However, two image sources are required which causes additional system cost.

<u>Depth cues</u> provide distance information from a single image source, but require knowledge (or assumptions) about the scenery. For example occlusion gives information about which object is closer. Further examples include the relative size of objects, or the absolute size of objects in the image plane, e.g., traffic signs with well-defined size and shape in automotive video systems. Also a texture gradient or light / shade falling onto an object provide depth information.

<u>Depth of focus</u> is widely used in digital cameras. The focus of a lens system is swept until the image contrast is maximized.

All of the above methods rely on feature extraction, which is a common requirement for all passive systems. Sufficient lighting as well as sufficient image contrast must be ensured to allow for image analysis. Probably the worst-case scenario for passive distance measurement techniques is a featureless, extended, plain surface. However, plain featureless walls are typically found on construction sites. Therefore, passive optical distance measurement techniques are inappropriate for distance measurement on a construction site.

2.1.2 Active Techniques

An active distance measurement system employs a light source for illumination of the target. The active illumination generates the missing features, for example, in form of a pattern or a light spot. Active distance measurement techniques can be classified by their principle of operation into geometrical distance measurement on the one hand and reflectometry on the other hand as shown in Fig. 2–2.

In <u>geometrical distance measurement</u>, the target distance is determined from the geometrical relationship of detectors, light source and the position of a lighting pattern on the target.

Fig. 2–2 (a) sketches a geometrical distance measurement by triangulation. A light source projects a feature, such as a point, line or pattern, at an angle onto a target. The target is imaged by a spatially resolving detector. This detector is separated from the light source by a known distance x (the baseline). The target distance is calculated from the position of the image of the feature in the detector plane. Because the accuracy of this method scales with detector resolution over focal length and the separation of sending and receiving path (the baseline x), the size of the system becomes impractical for millimeter resolution over tens of meters of range. Another straightforward but nevertheless quite fundamental characteristic of all triangulation based systems is the impossibility of a coaxial setup. A base is always needed.

Moreover, "shadowing" is a problem caused by the spatial separation of sending and receiving path. Light from the source has to reach the target and light from the target has to reach the receiver. Because of the separation of sending and receiving path, one of the two can be blocked by an obstacle. This is referred to as "shadowing". The requirement of high accuracy also increases the problem of shadowing because the accuracy increases with increasing separation of sending and receiving path,. Since obstacles are often present in measurement scenarios on a construction site, geometric distance measurement is not the appropriate choice here.

A special type of triangulation, namely pattern projection, has successfully been introduced into the mass market for 3D imaging and game control by PrimeSense in the Microsoft Kinect. The device covers a range of $0.8\ldots3.5\,\mathrm{m}$ and has a limited depth resolution of $10\,\mathrm{mm}$ at $2\,\mathrm{m}$ distance [10]. This performance, however, is not sufficient for the ranging application targeted here.

<u>Reflectometry</u> evaluates a difference of a signal sent to the target and reflected back from the target. In other words, reflectometry evaluates the echo of a signal. The echo arrives at the detector after the round-trip time of the signal to the target and back. Therefore, distance measurement by reflectometry can be referred to as time-of-flight (TOF) distance measurement.

In Fig. 2–2 (b) a light source sends out a signal (solid red lines) to a target at distance L. At the target some of the light is reflected or scattered back (dashed green lines). The detector receives some of the reflected signal as an echo. The target distance L is determined from the round-trip time of the signal to the target and back, and from the known propagation velocity of the signal. The distance resolution scales with the timing resolution of the system. As an alternative to directly measuring the round-trip time, a frequency or a phase difference between sent and received signal can be evaluated as will be discussed in Section 2.2.3.

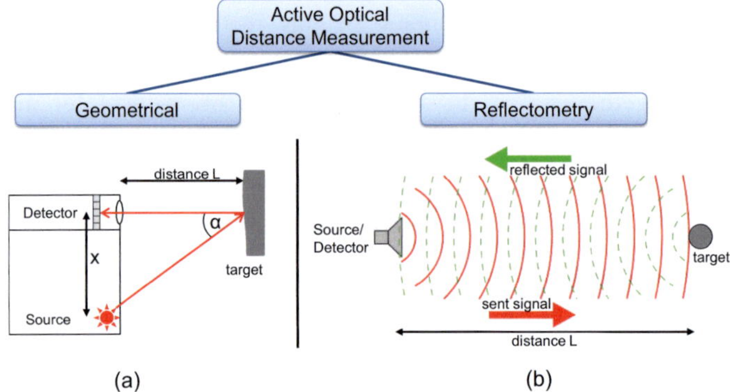

Fig. 2–2: Working principle of active optical distance measurement techniques
(a) Geometrical distance measurement by triangulation: A light source projects a beam onto the target. A spatially resolving detector is separated from the light source by x. The target distance L can be calculated from the position of the image of the beam in the detector plane, the projection angle α and the separation x (baseline).
(b) Reflectometry: A light source sends out a signal to a target at distance L. At the target some of the light is reflected or scattered backwards. The detector receives some of the reflected signal. The target distance L can be calculated from the time difference between sent and received signal and the known propagation velocity. (© graph (b) modified from [11])

An advantage of reflectometry over geometrical distance measurement is that a spatial separation of light source and detector does not form part of the measurement calibration. In reflectometry, source and detector can be closely spaced or could even be arranged as a coaxial system. Therefore, reflectometry enables a very compact system architecture that is ideal for a small-sized, handheld laser rangefinder. Furthermore, the limited number of components qualifies this approach for a low-cost laser rangefinder as will be shown in the next section.

2.2 Distance Measurement by Reflectometry

Generally speaking, reflectometry compares a difference of a signal and its echo, the reflected signal. In *time-domain reflectometry* (see Section 2.2.2), this difference is a time difference between sent and received signal that directly provides the time of flight. Therefore, time-domain reflectometry is also referred to as *direct* time-of-flight distance measurement. In *frequency-domain reflectometry* (see Section 2.2.3), said difference is a frequency or phase difference between sent and received signal. Frequency-domain reflectometry is also referred to as *indirect* time-of-flight distance measurement.

The concepts of optical time-domain reflectometry (OTDR) and optical frequency-domain reflectometry (OFDR) for distance measurement with a laser rangefinder find their equivalent in optical communication technology for fault localization in optical fibers [12]. Either the optical carrier itself [13] or a baseband signal [14] on the carrier can be modulated and evaluated for distance measurement. Different types of reflectometry in time- or frequency-domain are also known from radar applications in the microwave frequency range.

2.2.1 System Components

A time-of-flight distance measurement system comprises the basic components:

- Transmitter (abbreviated with Tx)
- Receiver   (abbreviated with Rx)
- Timing & Control Unit
- Human Machine Interface (HMI)

The transmitter comprises a light source, typically a laser diode (LD), with collimating optics and an electrical driver circuit for bias generation and direct modulation. The least expensive and most compact transmitters are implemented with directly modulated lasers [15]. This type of transmitter is therefore well-suited for a low-cost laser rangefinder. Fig. 2–3 (a) shows a graph of the optical output power P_{opt} of a laser diode as a function of the laser diode current I. Fig. 2–3 (b) shows a schematic of a directly modulated laser diode with bias tee. The laser diode is biased at a bias current I_{bias} above the lasing threshold current I_S with the bias tee. A capacitively coupled electrical signal modulates the laser diode current with I_{mod}. This modulation of the laser diode current translates into a modulation of the optical output power of the laser diode. Thus the electrical signal is intensity modulated onto the optical carrier.

A reference photodiode built into the laser diode package is used to control the optical output power to ensure a stable operating point and eye-safe operation.

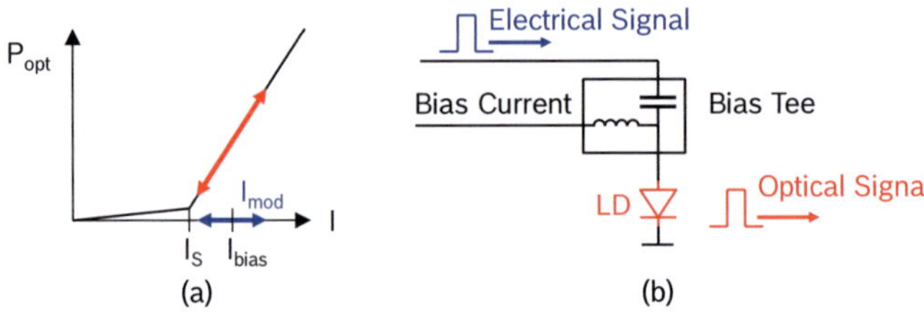

Fig. 2–3: Transmitter (a) optical output power as a function of the current of a laser diode (b) directly modulated laser diode with bias tee.

An optical carrier wavelength λ_0 of typically 635 nm or 650 nm is used because a visible beam is required in our application and inexpensive laser diodes are available at these wavelengths. The optical frequency of the carrier is $f_0 = c / \lambda_0 = \omega_0 / 2\pi$, wherein c denotes the vacuum speed of light and ω_0 denotes the angular frequency. Free space propagation in air is approximated with vacuum permittivity with $\varepsilon_r = 1$ and vacuum permeability with $\mu_r = 1$. The refractive index is $n_{\text{air}} = \sqrt{\varepsilon_r \mu_r} = 1$ and can therefore be omitted in some formulae. Air is considered an ideal dispersion-free medium throughout this thesis.

The receiver comprises a photodetector for direct detection, typically an avalanche photodiode (APD), a band-pass filter centered at the optical carrier frequency at f_0 for suppression of background light, and the receiver optics. A fixed focus receiver lens is used for low system cost.

The timing and control unit (TCU) controls the transmitter and processes the received signal. The TCU generates the electrical driving signals for a desired intensity modulation. Moreover, the difference between sent and received signal is determined and the target distance calculated therefrom. Two different approaches for distance measurement, time-domain reflectometry and frequency-domain reflectometry, will be described in the subsections below.

The Human Machine Interface (HMI) displays the measured values and provides the user interface for device control, e.g., a keyboard or a touchscreen.

2.2.2 Time-Domain Reflectometry

Time-domain reflectometry, also referred to as direct time-of-flight distance measurement in this context, is the straightforward approach. In radar systems, time-domain reflectometry corresponds to an impulse radar. The same concept, known from radar systems that operate at microwave frequencies, is transferred to the optical part of the spectrum for a laser rangefinder. A single pulse is intensity modulated onto the optical carrier. The electrical modulation signal that causes the optical pulse also starts a timer at time t_{start}. The optical pulse propagates towards the target and some of the backscattered light travels back to the receiver. The received signal stops the timer at time t_{stop}. Fig. 2–4 illustrates this principle. The round-trip time or time of flight τ is the difference between t_{stop} and t_{start},

$$\tau = t_{stop} - t_{start} .\tag{2.1}$$

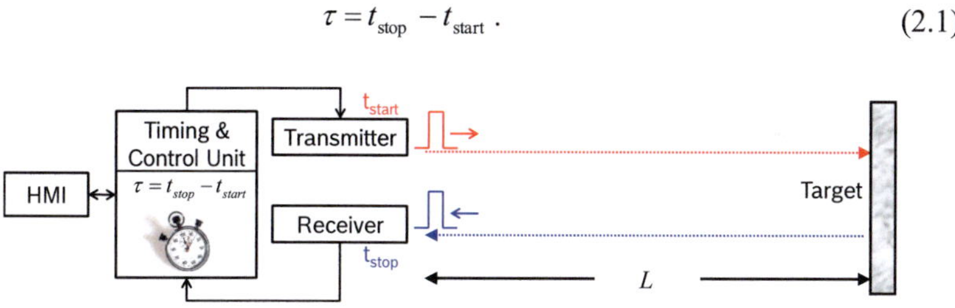

Fig. 2–4: Distance measurement using time-domain reflectometry. The transmitter sends a light pulse (red) to the target at time t_{start}. Some of the light is backscattered (blue) at the target and received by the receiver at time t_{stop}. The timing & control unit calculates the target distance L from the time of flight $\tau = t_{stop} - t_{start}$ and displays the result on the human machine interface (HMI).

The propagation speed of the signal is given by the group velocity v_g. Since air is approximated as a dispersion-free medium with vacuum refractive index, the pulse travels towards the target and back at the vacuum speed of light c. The target distance L results from the measured time of flight τ and the speed of light c. A factor of $1/2$ accounts for way and return,

$$L = \frac{1}{2} c\tau .\tag{2.2}$$

Correspondingly, the distance measurement error or the distance resolution ΔL depends on the measurement accuracy or timing resolution $\Delta\tau$, i.e., on how fine the time of flight τ can be resolved,

$$\Delta L = \frac{1}{2}c\Delta\tau . \tag{2.3}$$

Freude shows in [12] that the distance resolution ΔL is proportional to the sum of the observation period of the receiver T_{Rx} and the pulse duration T_{pulse}, $\Delta L \sim \left(T_{Rx} + T_{\text{pulse}}\right)$. Thus, the observation period and pulse duration should be short for a high distance resolution. A high timing resolution, however, corresponds to a large bandwidth $\Delta f \sim 1/\left(T_{Rx} + T_{\text{pulse}}\right)$. The distance resolution in a pulsed system thus scales with the inverse of the receiver bandwidth, $\Delta L \sim 1/\Delta f$.

Probably the shortest pulse is a single oscillation of the electric field of the optical carrier [16]. In a laboratory setup, a short pulse duration or a short observation period of the receiver is not the limiting factor for millimeter-precision distance measurement. However, the generation requires complex laboratory setups. The system cost definitely exceeds the several tens of Euros that a craftsman or do-it-yourselfer would be willing to spend on a measuring device for a construction site.

For the given application a distance resolution of 1 mm corresponds to 6.7 ps timing resolution. A sampling scope for a laboratory setup would require a 150 GHz sampling rate to resolve 6.7 ps steps. Moreover, distances from 5 cm to tens of meters require good linearity of the timing circuit over the entire measurement range.

For the target low-cost laser rangefinder, system components with a large bandwidth are not attractive because of the typically higher cost of large bandwidth components.

2.2.3 Frequency-Domain Reflectometry

As an alternative to time-domain reflectometry, the target distance can also be determined with a measurement system that operates in the frequency domain. Instead of transmitting individual impulses, the signal is a continuous wave (CW). In frequency domain, the target distance can be determined based on the difference of the frequency or the phase of sent and received signal.

2.2.3.1 Frequency difference

This type of frequency-domain reflectometry is generally known from frequency-modulated continuous wave (FMCW) radar systems. In an FMCW system, the transmitter (Tx) sends out a frequency modulated signal to the target, for example a saw-tooth frequency sweep as shown in Fig. 2–5. The time of flight τ can be calculated from the known slope s (in Hz/s) of the frequency sweep and the frequency difference f_d between sent (f_{Tx}) and received (f_{Rx}) signal,

$$\tau = \frac{f_{Rx} - f_{Tx}}{s} = \frac{f_d}{s}. \tag{2.4}$$

Using Eq. (2.2), the target distance L is given by

$$L = \frac{1}{2} c \frac{f_d}{s}. \tag{2.5}$$

The distance resolution in an FMCW-system scales with the frequency resolution, $\Delta L \sim \Delta f_d$.

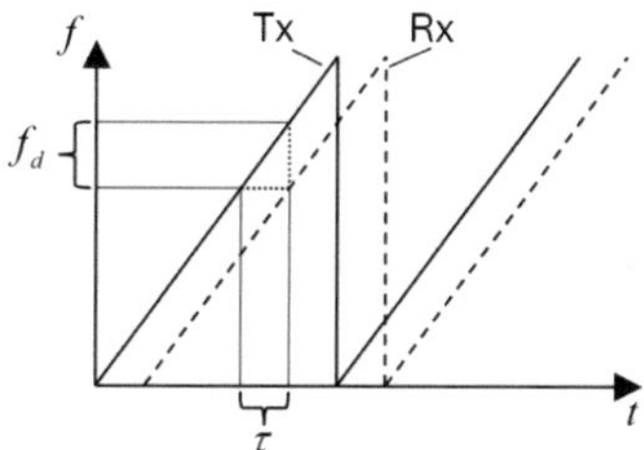

Fig. 2–5: Frequency-time diagram for a frequency-modulated continuous wave (FMCW) distance measurement system using a saw-tooth frequency sweep. The time of flight τ can be determined from the frequency difference f_d between sent (Tx) and received (Rx) signal.

An FMCW-system has the advantage that the slope can be adjusted for a given application. For example, a shallow slope can be used for measuring a long target distance with lower distance resolution, whereas a steep slope can be used for measuring a short target distance with higher distance resolution. Optical frequency modulation or baseband frequency modulation can be used for FMCW distance measurement.

An FMCW-system, however, requires a transmitter that generates a frequency sweep with high precision. Any non-linearity of the ramp impairs the distance measurement. For a low-cost system, it would be advantageous to have a much simpler modulator, for example only operating at a limited number of discrete frequencies or even at a single modulation frequency.

2.2.3.2 Phase difference

Instead of determining a frequency difference between sent and received signal, the phase difference between sent and received signal at a single frequency can be evaluated.

This principle is well known from interferometry at optical wavelengths. Optical interferometry enables distance measurements with high resolution on the optical wavelength scale by interference of light reflected from the target and light from a reference arm of the interferometer. The distance is determined from the optical phase, thus the phase of the electric field. System complexity, cost and alignment are adequate for high precision instrumentation, but are not feasible for a low-cost system that has to withstand the harsh environment of a construction site.

However, the principle of interferometry cannot only be applied to interfering optical signals but can equally be used by comparing the envelope of a modulated optical signal after detection with a reference signal in the electrical domain. Thus the phase shift between the envelope of the received optical signal and the reference, for example the electrical modulation signal of the transmitter, can be evaluated to determine the target distance. The setup does not require an optical beam splitter and careful alignment of the reference path. Instead, the optical setup reduces to a light source and detector with associated optics. Hence frequency-domain reflectometry also qualifies for a low-cost, handheld laser rangefinder.

Fig. 2–6 illustrates the principle of frequency domain reflectometry by evaluating a phase difference between sent and received signal. The transmitter sinusoidally modulates the intensity I_{Tx} of the optical carrier,

$$I_{Tx}(t) = I_1\left(1 + \cos(\omega_m t + \Phi_{Tx})\right). \tag{2.6}$$

After having propagated to the target and back to the receiver, the receiver detects the sum of the backscattered modulated signal and background light with intensity I_{BL},

$$I_{Rx}(t) = I_2\left(1 + \cos(\omega_m t + \Phi_{Rx})\right) + I_{BL} \,. \tag{2.7}$$

It should be noted that the received intensity I_2 provides only limited distance information. A poorly reflective target at short distance can yield the same intensity as an ideal target further away.

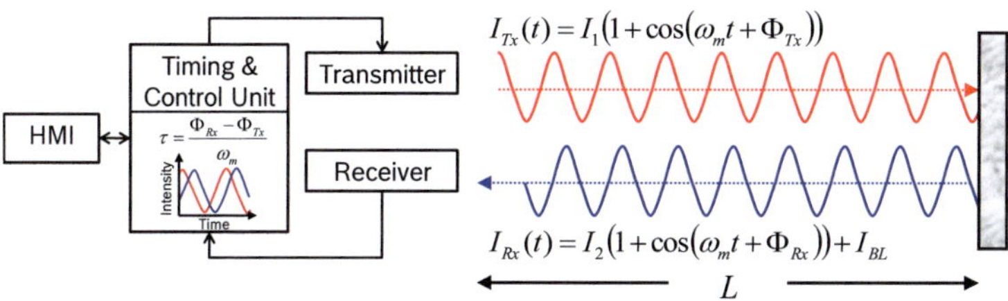

Fig. 2–6: Distance measurement using frequency-domain reflectometry and evaluating the phase difference between sent and received signal. The transmitter sends continuous sinusoidally intensity modulation signal I_{Tx} (red) to the target. Some of the light is backscattered (blue) at the target and received by the receiver together with background light I_{BL}. The timing & control unit calculates the target distance L via the time of flight τ from the phase difference, $\varphi = \Phi_{Rx} - \Phi_{Tx}$ and the angular modulation frequency ω_m. The result is displayed on the human machine interface (HMI).

The receiver phase Φ_{Rx} and transmitter phase Φ_{Tx} are linked by the time of flight τ and the angular modulation frequency $\omega_m = 2\pi f_m$. Thus the phase difference φ between sent and received signal is given by

$$\varphi = \Phi_{Rx} - \Phi_{Tx} = \omega_m \tau \,. \tag{2.8}$$

Analogous to Eq. (2.2), the target distance L can now be calculated from the time of flight, whereas the time of flight τ is indirectly represented by the phase difference φ between sent and received signal,

$$L = \frac{1}{2}c\tau = \frac{1}{2}c\,\frac{\varphi}{\omega_m} \,. \tag{2.9}$$

The distance resolution ΔL now depends on the phase resolution $\Delta \varphi$ and scales with the modulation frequency $f_m = \omega_m / 2\pi$,

$$\Delta L = \frac{1}{2}c\Delta\tau = \frac{1}{2}c\,\frac{\Delta\varphi}{\omega_m} \,. \tag{2.10}$$

A high modulation frequency f_m relaxes the phase accuracy requirement. To achieve 1 mm distance resolution ΔL at a modulation frequency $f_m = 10\,\mathrm{MHz}$, the phase must be known with $0.024°$ accuracy. At a modulation frequency $f_m = 1\,\mathrm{GHz}$, the required phase accuracy is relaxed to $2.4°$. For a measurement system with a given phase accuracy, the modulation frequency should be as high as possible.

2.2.3.3 Distance ambiguity

Measured phase difference φ to target distance L is an ambiguous relationship. Multiple target distances $L(k)$ result in the same phase difference φ (ranging from 0 to 2π),

$$L(k) = \frac{1}{2}c\tau = \frac{1}{2}c\left(\frac{\varphi + 2\pi k}{\omega_m}\right) \quad , \quad k = 0, 1, 2, \ldots \tag{2.11}$$

The range over which there is no ambiguity is called the unambiguity range L_U given by half the modulation wavelength $\lambda_m = c / f_m$,

$$L_U = \frac{1}{2}\frac{c}{f_m} = \frac{1}{2}\lambda_m . \tag{2.12}$$

Thus a low modulation frequency provides a long unambiguity range. However, as described above, a high modulation frequency is desired for a low distance measurement error in a measurement system with a limited phase accuracy. The solution to these conflicting interests is measuring at least two closely spaced high frequencies f_{m1}, f_{m2} [17]. The unambiguity range L_U can now be calculated from the beat frequency $f_{m2} - f_{m1}$,

$$L_U = \frac{1}{2}\frac{c}{f_{m2} - f_{m1}} . \tag{2.13}$$

For example, a desired unambiguous range of 100 m leads to a frequency spacing of 1.5 MHz, e.g., modulating the carrier at a first modulation frequency $f_{m1} = 1.0000\,\mathrm{GHz}$ and then at a second modulation frequency $f_{m2} = 1.0015\,\mathrm{GHz}$. Even though multiple modulation frequencies are used, in each measurement the power is confined to a narrow spectral range and thus a high power spectral density can be reached. Each measurement is a narrowband measurement. Overall, a frequency-domain reflectometer can be considered a narrowband system – with the respective cost benefit for system components.

Moreover, CW systems do not require short-pulse lasers with high peak power, but can use low-cost, off-the-shelf laser diodes.

2.2.3.4 Receivers for Frequency-Domain Reflectometry

As mentioned above (see Section 2.2.1), the receiver comprises a photodetector for direct detection of the received signal at low cost. In order to reduce the frequency requirements for an analog-to-digital (A/D) converter that digitizes the received signal for further processing in the digital domain, it is advantageous to downconvert the received signal before A/D conversion. The received signal is mixed with a local oscillator at frequency ω_{LO} after detection in the electrical domain. Fig. 2–7 shows a photodetector with a subsequent pair of mixers.

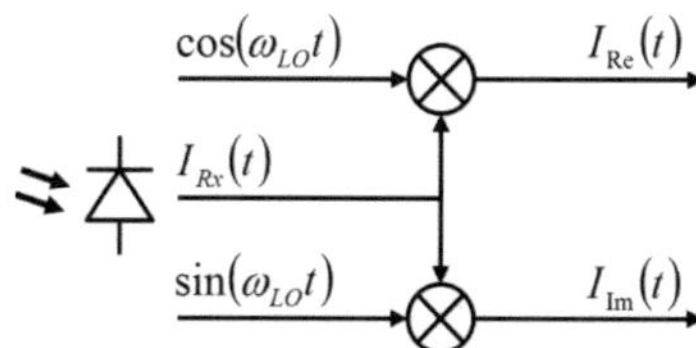

Fig. 2–7: Photodetector with subsequent pair of mixers. Received signal intensity $I_{Rx}(t)$, mixer frequency ω_{LO}, outputs of upper mixer arm $I_{Re}(t)$ and lower mixer arm $I_{Im}(t)$.

Referring to Eq. (2.7), the received signal intensity is given by

$$I_{Rx}(t) = I_2\left(1 + \cos(\omega_m t + \Phi_{Rx})\right) + I_{BL} \,. \tag{2.14}$$

The output of the upper multiplicative mixer is thus given by

$$\begin{aligned}
I_{Re}(t) &= \left(I_2\left(1 + \cos(\omega_m t + \Phi_{Rx})\right) + I_{BL}\right)\cos(\omega_{LO}t) \\
&= \left(I_2 + I_{BL}\right)\cos(\omega_{LO}t) \\
&\quad + \frac{1}{2}I_2\cos\left((\omega_m + \omega_{LO})t + \Phi_{Rx}\right) \\
&\quad + \frac{1}{2}I_2\cos\left((\omega_m - \omega_{LO})t + \Phi_{Rx}\right).
\end{aligned} \tag{2.15}$$

The terms at ω_{LO} and at the sum frequency $\omega_m + \omega_{LO}$ can be filtered out such that the remaining terms in the upper and lower mixer arms are

$$I_{\mathrm{Re},2}(t)=\frac{1}{2}I_2\cos\left((\omega_m-\omega_{LO})t+\Phi_{Rx}\right)\text{ and}$$

$$I_{\mathrm{Im},2}(t)=\frac{1}{2}I_2\sin\left((\omega_m-\omega_{LO})t+\Phi_{Rx}\right).$$

(2.16)

It should be noted that the phase Φ_{Rx} of the received signal is maintained throughout the mixing process and can thus be evaluated for determining the phase difference between sent and received signal as described above in Section 2.2.3.2. For the case that the transmitter phase $\Phi_{Tx}=0$, the phase difference between sent and received signal is $\varphi=\Phi_{Rx}$. A phase in this context again refers to the phase of the envelope of the intensity modulated optical carrier, not an optical phase.

If the local-oscillator frequency is selected to coincide with the modulation frequency, so that $\omega_m=\omega_{LO}$, one speaks of *homodyne* detection.

In the case of *heterodyne* detection the local oscillator is chosen to differ from the modulation frequency, $\omega_m\neq\omega_{LO}$, such that the intermediate frequency $\omega_m-\omega_{LO}$ is, for example, in the kHz region. Analog to digital converters for sampling DC or kHz signals are commercially available in off-the-shelf microcontrollers at low cost.

The book by Hugenschmidt [18] provides further information on direct and indirect time-of-flight laser distance measurement. Further information about homodyne and heterodyne detection in the context of optical communication systems can be found in [15].

2.3 State-of-the-Art Ranging Systems

Firstly, the operating principle of the current Bosch laser rangefinders GLM 250 and GLM 80 [4, 5] is outlined (see Section 2.3.1). These rangefinders are commercially available and manufactured on an industrial scale. They are considered 'semi-integrated' ranging systems since a part of the system components for determining the time-of-flight is integrated into an application-specific integrated circuit (ASIC). Sections 2.3.2 and 2.3.3 give an overview of integrated ranging systems that have successfully been demonstrated and published in recent years. These systems are considered 'integrated' ranging systems since they further enable the integration of the detector into one ASIC along with the timing and control circuitry. The integration of the light source into the ASIC is not considered at this stage. None of the integrated ranging systems achieves the performance required for the present application yet, however they show promising approaches. A performance summary table is provided at the end of this Section.

2.3.1 State-of-the-Art Bosch Laser Rangefinder & Limitations

The current Bosch laser rangefinders GLM 250 and GLM 80 [4, 5] employ frequency-domain reflectometry for distance measurement and evaluate a phase difference between sent and received signal. Photographs of these devices have been shown in Fig. 1–3. Fig. 2–8 gives an overview of the core elements.

Operating Principle

A quartz oscillator provides a stable 8 MHz reference frequency for a microcontroller (µC) and an application-specific integrated circuit (ASIC). A phase locked loop (PLL) in that ASIC generates the radio frequencies for directly modulating the laser diode current with $f_m = 2\pi\omega_m$ and the gain of an avalanche photodiode (APD) with $f_{LO} = 2\pi\omega_{LO}$. The two frequencies differ by some kHz. Since only the frequency generation circuits, modulators, drivers and some control electronic are integrated in the ASIC, the system is referred to as a semi-integrated ranging system. The light source and the detector are not integrated in the ASIC. The angular modulation frequency ω_m is capacitively coupled to the laser diode, so that the laser diode current is modulated. An intensity modulated beam of light is sent out to the target through a collimator. The ASIC also controls the bias current and ensures an eye-safe average optical output power of the laser diode of 1 mW.

The receiving lens collects both background light and signal light backscattered from the target. An optical band-pass filter (OBPF) centered at the optical wavelength λ_0 of the laser suppresses most of the background light. Unfortunately for the signal-to-noise ratio, a fair amount of background light is still transmitted because of the limited quality of the low-cost OBPF, and because the filter bandwidth has to account for some nanometers laser diode wavelength drift over temperature. The avalanche photodiode is reverse biased at ~120V. The angular modulation frequency ω_{LO} is capacitively coupled to the avalanche photodiode to modulate the APD gain. Instead of detecting the signal and then mixing the received signal with the local oscillator at ω_{LO}, the gain-modulated APD already acts as a heterodyne receiver as described above in Section 2.2.3.4. The APD is followed by an electrical band-pass filter centered at the intermediate frequency $\omega_m - \omega_{LO}$ which blocks the high frequency components from the mixing process and also suppresses any offset such that the intermediate frequency remains. The filtered signal is amplified and then digitized with the A/D converter (ADC) of a low-cost microcontroller. The phase difference between sent and received signal is evaluated to determine the target distance. The result is displayed on a human machine interface (HMI).

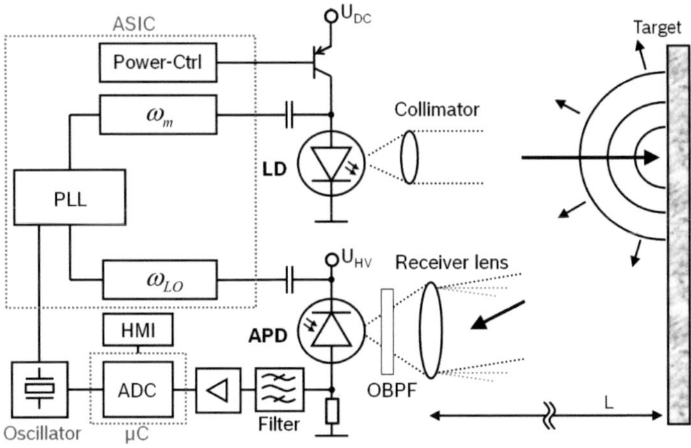

Fig. 2–8: State-of-the-art laser rangefinder system based on frequency-domain time-of-flight distance measurement. The receiving avalanche photodiode (APD) acts as an electrical heterodyne mixer. Application-specific integrated circuit (ASIC), phase locked loop (PLL), laser diode (LD), optical band-pass filter (OBPF), analog-to-digital converter (ADC), human machine interface (HMI), angular modulation frequency of the transmit path ω_m, angular modulation frequency of the receiver path ω_{LO}.

The current APD-based system achieves a distance measurement performance (see Table 1.1) well-sufficient for the requirements on a construction site. However, the system is built up from costly components and has limitations that motivate system integration.

Limitations

Whilst achieving a performance that already exceeds the requirements, there are drawbacks of the state-of-the-art system which leave room for improvement:

Sensitivity: Large aperture optics are required to gather enough light for a good overall signal-to-noise ratio (Fig. 2–9 (a)). A detector with high sensitivity would be advantageous to shift the burden from optics to electronics and allow for a smaller optical system. The optical system and operating conditions of a handheld laser rangefinder will be discussed in detail in Chapter 5.

Cost: CMOS integration instead of a discrete expensive APD and associated electronics including transimpedance amplifier and filter (Fig. 2–9 (b)).

Size: The current system uses a 'reference flap' (Fig. 2–9 (c)) to compensate for thermal laser diode drift. The reference flap mechanically switches between sending the laser beam onto the target or through a device-internal reference path for calibration. In a semiconductor technology with an integrated photodetector, a second photodetector can be integrated and optically separated [19]. One photodetector can be used for target distance measurement and the other one for parallel internal reference measurement on one chip [20]. The bulky reference flap driven by a magnetic coil becomes obsolete.

Alignment in manufacturing: Indirect cost and size benefits open up as the alignment requirements can be relaxed. Besides two detectors – one for light from the target, one for a device-internal reference – further detectors can be implemented on chip to compensate for manufacturing inaccuracies. An extended two-dimensional array of detectors can be implemented. Instead of aligning the optics with micrometer precision onto the light sensitive active area of a single detector, one can, without any alignment, simply choose the detector out of the array of detectors that is at the correct position [21, 22].

Alignment stability during device lifetime: An industrial-grade laser rangefinder must endure temperature variations and drops on a concrete floor while maintaining micrometer alignment. If an extended two-dimensional array of integrated detectors is available, a misalignment or changes of the alignment over device lifetime are not problematic. An intelligent readout system can detect when the laser spot is out of alignment, and simply evaluate the signal from a different detector out of the array of detectors that then receives the signal light scattered back from the target [21, 22].

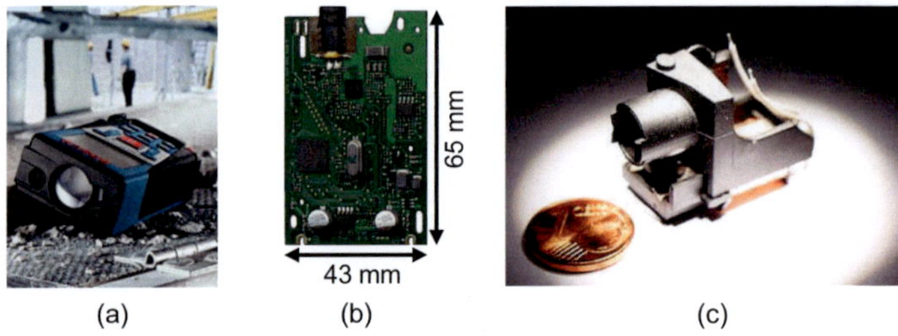

(a)   (b)   (c)

Fig. 2–9: Conventional laser rangefinder. (a) Bosch GLM 250 with 25 mm diameter high quality glass lens, (b) Bosch DLE 50 printed circuit board with RF-ASIC and separate avalanche photodiode (c) Bosch DLE 50 mechanical reference flap with magnetic coil for device-internal reference distance measurement (courtesy of Robert Bosch GmbH)

2.3.2 Ranging Systems using Integrated Modulated Pixels

Instead of modulating the gain of an external detector the straightforward approach is modulating the gain of an integrated detector that is fabricated along with the timing and control circuitry in the same semiconductor process. A heterodyne mixing CMOS-integrated APD has been presented in [23] with a distance resolution of 42 mm. This value does not account for temperature drifts or background light.

Even better performances have been reached with modulated CMOS/CCD pixels, so-called lock-in CCDs. CCDs (charge coupled devices) are the most commonly used image sensors in digital cameras. Starting with ±10 cm resolution in the late 1990s [24, 25], this technology is now commercially available in 3D camera systems by Swiss Ranger [26] and PMDtec [27]. The systems achieve some millimeters accuracy (details described below; see also Table 2.1). The technology will be employed in the automotive market in the near future.

In brief, in a conventional CCD image sensor photo-generated electrons and holes (eh-pairs) are separated by an electric field. The photo-generated) electrons are accumulated in one storage node per pixel. The photo-generated holes are collected by a common substrate contact. The amount of photo-generated charges per time interval represents the incident light intensity of that pixel. A modulated pixel for distance measurement features a plurality of storage nodes per pixel. The structure and operating principle of accumulating charges in two different storage nodes is shown in Fig. 2–10. The positively biased, transparent electrodes U_0+U_{mod} and U_0-U_{mod} of a pixel are modulated in synchronism with the intensity modulation of the light source. The applied voltage alters the electric field and thereby controls the potential curve and movement of photo-generated electrons (illustrated by a red dot) to the storage nodes A_1 and A_3. The accumulated charge is read out via $U_{out,1}$ and $U_{out,3}$. The photo-generated holes (not shown) are collected by a common substrate contact as described before.

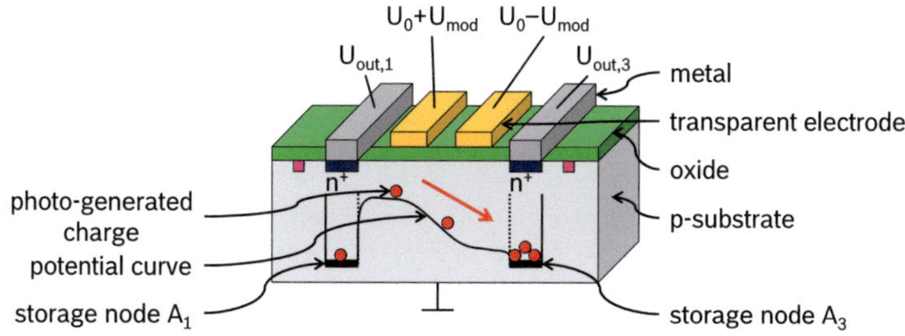

Fig. 2–10: Integrated modulated pixel [28]. Combined graph of semiconductor layers (metal, transparent electrode, oxide, and substrate) and potential curve in substrate. The potential curve causes a majority of photogenerated charges (red dots, representing electrons) to move towards storage node A_3. The photogenerated holes are collected by a common substrate contact.

Fig. 2–11 shows two different potential curves between the storage nodes A_1 and A_3. In the upper row, U_0+U_{mod} is applied at the left transparent electrode and U_0-U_{mod} at the right transparent electrode (see Fig. 2–10). The potential between the storage nodes decreases from the left to the right. This setting is applied during a first half $\Delta t_{mod,1}$ of the modulation period of the light source. In the lower row in Fig. 2–11, U_0-U_{mod} is applied at the left transparent electrode and U_0+U_{mod} at the right transparent electrode. The potential between the storage nodes increases from the left to the right. This setting is applied during a second half $\Delta t_{mod,2}$ of the modulation period of the light source.

Fig. 2–11 (a) illustrates the accumulation of photo-generated charges in storage nodes A_1 and A_3 for unmodulated light. A photon (illustrated by a yellow dot) is absorbed in the detector and creates an electron-hole pair. A photo-generated electron is illustrated by a red dot. The potential profile during $\Delta t_{mod,1}$ causes a movement of the photo-generated charge towards storage node A_3. Analogously, the potential profile during $\Delta t_{mod,2}$ causes a movement of the photo-generated charge towards storage node A_1. Since the light in Fig. 2–11 (a) is not modulated, the same average number of photons can be expected during $\Delta t_{\mathrm{mod},1}$ and $\Delta t_{\mathrm{mod},2}$. Thus, the storage nodes contain the same average number of charge carriers after an entire modulation period.

Fig. 2–11 (b) illustrates the accumulation in storage nodes A_1 and A_3 for modulated light. The potential curve is modulated in synchronism with the light source. In this simplified example, two photons are incident during $\Delta t_{\mathrm{mod},1}$. The photo-generated charges are accumulated in storage node A_3. During $\Delta t_{\mathrm{mod},1}$, no photons are detected and hence no photogenerated charges accumulated in storage node A_1. Thus, after an entire modulation period, the charges in the storage nodes represent the intensity of the modulated light during $\Delta t_{\mathrm{mod},1}$ and $\Delta t_{\mathrm{mod},2}$.

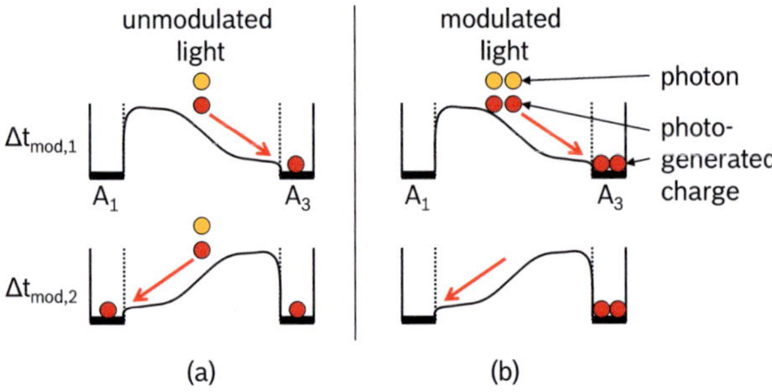

Fig. 2–11: Two phases of operation of a modulated pixel [28] (a) with unmodulated background light and (b) with modulated light, wherein the potential curve is modulated in synchronism with the light source. Yellow dots: photons. Red dots: electrons.

For the case of four storage nodes per pixel, Fig. 2–12 shows a graph of accumulated charges per storage node (A_0 to A_3) over time. Designs for modulated pixels with four storage nodes are disclosed in a patent by Schwarte [29]. When viewed from the top, the modulated pixel is a two-dimensional structure, for example a square. A storage node can be implemented at each of the four sides. A pixel with four storage nodes (only two storage nodes are shown in Fig. 2–10 and Fig. 2–11) can be operated such that the photo-generated charges are accumulated in the first storage node A_0 during the first quarter of the modulation period, onto the second storage node A_1 during the second quarter of the modulation period and so on. This effectively samples the incident wave with four samples per period (Fig. 2–12). Further details about sampling a periodic signal are provided in Appendix C.

Offset, phase and amplitude of the sampled received intensity can be calculated using the Fourier transform, wherein X_l denotes a value in the frequency domain and x_k denotes the measured sample values of the storage nodes $A_0 \ldots A_3$,

$$X_l = \frac{1}{N} \sum_{k=0}^{N-1} x_k e^{j2\pi l \frac{k}{N}} \qquad l = 0, \ldots, N-1. \tag{2.17}$$

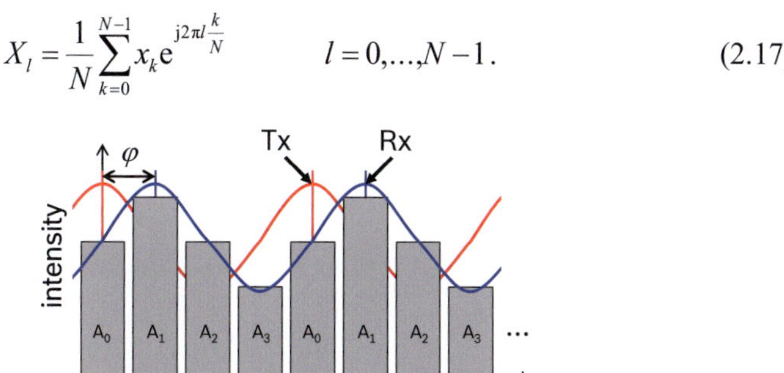

Fig. 2–12: Sampling of the incident light in synchronism with the modulation of the light source. The red curve denotes the modulated intensity emitted from the light source (Tx); the blue curve denotes the intensity received at the modulated pixel (Rx). The bars represent the number of charges per storage node (A_0, A_1, A_2, A_3) of the modulated pixel. The phase difference between Tx and Rx is φ.

The DC offset is given by the 1st element of the Fourier transform X_0, the fundamental modulation frequency phase and amplitude by the 2nd complex element X_1. Other elements can be discarded if higher harmonics are of no interest. The target distance is calculated using Eq. (2.9) from the phase of the fundamental frequency,

$$\begin{aligned}
X_1 &= \frac{1}{4} \sum_{k=0}^{3} x_k e^{j2\pi \frac{k}{N}} \\
&= \frac{1}{4} \left(x_0 e^{j2\pi \frac{0}{4}} + x_1 e^{j2\pi \frac{1}{4}} + x_2 e^{j2\pi \frac{2}{4}} + x_3 e^{j2\pi \frac{3}{4}} \right) \\
&= \frac{1}{4} \left([x_0 - x_2] + j[x_1 - x_3] \right).
\end{aligned} \tag{2.18}$$

With the real part $\Re(X_1) = \frac{1}{4}[x_0 - x_2]$ and the imaginary part $\Im(X_1) = \frac{1}{4}[x_1 - x_3]$, the phase of the fundamental modulation frequency for the case of four samples can be directly determined from the sample values as

$$\varphi = \arctan\left(\frac{x_1 - x_3}{x_0 - x_2}\right). \tag{2.19}$$

It should be noted that charges can be integrated over several RF periods before readout to reduce readout speed requirements. A detailed study of integrated modulated pixels can be found in the PhD thesis by Lange [30].

A drawback of CCD technology is that the fabrication is more expensive than standard CMOS. Specialized dedicated fabrication processes for ranging applications can optimize performance at additional cost. Manufacturing reliability and yield is limited [31] and will require further effort before mass production can be entered.

For a laser rangefinder, distance accuracy is crucial. This rules out systems which suffer from several millimeters temperature drift of the measured distance. In other words, a system that is capable of determining the distance with a repcatability error in the sub-millimeter range is still not adequate for a laser rangefinder, if the distance offset is not known. In spite of a device-internal reference path [31] the SR4000 suffers from a maximum temperature drift of the measured distance of $1.5\,\text{mm/°C}$ ($0.5\,\text{mm/°C}$ around room temperature) [26]. The internal reference path is implemented as a light guide of known length from the light source to a part of the detector.

The measurement range and accuracy of the SR4000 depend on the modulation frequency. The SR4000 operates at a single modulation frequency of $30\,\text{MHz}$ ($15\,\text{MHz}$), which corresponds to a maximum target distance of $L = 5\,\text{m}$ ($L = 10\,\text{m}$). Definitions of the distance measurement accuracy and of the distance measurement errors used below are provided in Appendix A. The repeatability error at $2\,\text{m}$ target distance is $\sigma_{2\text{m}} = 4\,\text{mm}$ ($\sigma_{2\text{m}} = 6\,\text{mm}$). Since the measurement accuracy comprises twice the repeatability error, the worst-case measurement accuracy at $2\,\text{m}$ target distance is at least $\varepsilon(2\text{m}) \geq 2\sigma_{2\text{m}} = 8\,\text{mm}$ ($\varepsilon(2\text{m}) \geq 2\sigma_{2\text{m}} = 12\,\text{mm}$). It should be noted that the worst-case systematic distance error ε_μ, for example due to the temperature drift of the measured distance, still has to be added on top of this value. This distance measurement accuracy is not sufficient for the application targeted here (see Table 1.1).

Going to higher modulation frequencies is expected to further improve the performance. However, the modulation frequency for this type of detector cannot be arbitrarily increased. At a high modulation frequency, a modulated pixel is not capable of clearly separating the charges into different storage nodes any more. Thus, the demodulation contrast decreases. Therefore, the modulation frequency of modulated pixel should not exceed some tens of MHz. For example the modulated pixel in [32] is operated at $f_m = 20\,\text{MHz}$ only.

2.3.3 Ranging Systems using Integrated SPADs

A single-photon avalanche diode (SPAD) is an alternative type of photo-sensor for use in a ranging system. SPADs are attractive because they are compatible with existing low-cost semiconductor fabrication processes. SPADs have successfully been demonstrated in a CMOS process for low-cost image sensors [33]. This process is used on an industrial scale in mass production of cell phone cameras.

A single-photon avalanche diode (SPAD) is effectively a triggered photonic device that generates a digital output pulse when struck by a photon. SPADs will be discussed in detail in Chapter 3. The output signals of a SPAD are impulses of constant height and constant width. This digital output behavior makes the device well suited for subsequent digital signal processing with either time-domain or frequency-domain distance measurement (see Sections 2.2.2 and 2.2.3).

2.3.3.1 Time-Domain Reflectometry with SPADs

A method for time-domain or direct time-of-flight (TOF) distance measurement with single-photon detectors is called time-correlated single-photon counting (TCSPC). TCSPC builds a histogram of the arrival times of photons. The histogram denotes the number of photons detected in a time slot or sampling window (bin) after a start trigger at time zero when a laser pulse is sent to the target. TCSPC is known from low-light level, time-correlating applications such as fluorescence lifetime imaging [34] in biology. The associated timing circuits are an entire research area of its own. An overview of timing circuits is given in [35].

Fig. 2–13 shows a histogram of the expected number of photons per bin. The background light is constant and each bin has an identical width $b_k = b$, wherein k is the bin number. Therefore, the same expected number of background light photons is incident in each bin of the histogram. At time $t_{start} = 0$ a short laser pulse is sent to the target. Some of the light is scattered back to the receiver. At time $t_{stop} = t_0$ the photons from the laser pulse impinge on the SPAD. The photons received from this laser pulse are represented by a peak in the histogram. The distance can be directly calculated from the time of flight $\tau = t_{stop} - t_{start} = t$ using Eq. (2.2).

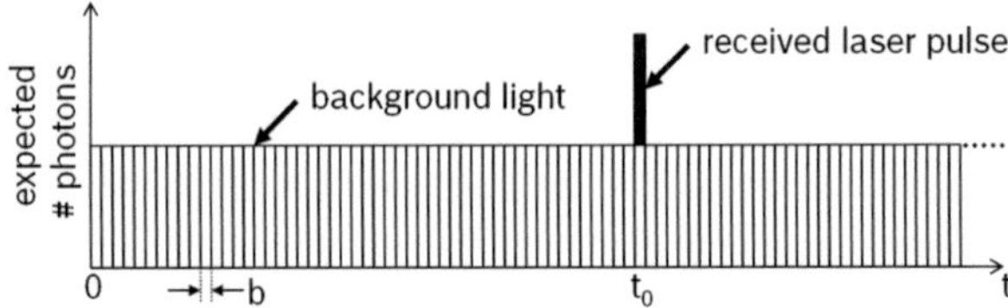

Fig. 2–13: Histogram of expected number of photons incident on a single-photon avalanche diode (SPAD) during different time slots (bins) of identical bin width b. At time $t_{start} = 0$ a laser pulse is sent to the target. At time $t_{stop} = t_0$ the photons from the laser pulse are incident on the SPAD. The target distance can be calculated from the time of flight $\tau = t_{stop} - t_{start} = t_0$.

The histogram in Fig. 2–13 is idealized insofar as all time slots (bins) have the same width $b_k = b$ and are equidistantly distributed in time. Fig. 2–14 (a) illustrates this idealized case. Real-world implementations for measuring such histograms, however, suffer from non-idealities of the bin widths and bin positions as illustrated in Fig. 2–14 (b). Technical implementations of a binning architecture for generating bins are described in more detail in Section 5.3.3.

The bin widths of the non-ideal binning are given by b'_k. If the bins follow directly adjacent to each other, the bin width also affects the position of the center of the bins, t_k for an ideal binning and t'_k for a non-ideal binning, by

$$t_k = kb + \frac{1}{2}b$$

$$t'_k = \frac{1}{2}b'_k + \sum_{l=0}^{k-1} b'_l \,.$$

(2.20)

The difference between actual bin width and ideal bin width of bin number k is referred to as the differential nonlinearity DNL_k. The difference INL_k between actual bin position t'_k (red) and ideal bin position t_k (blue) is indicated in Fig. 2–

14 (b). Since the actual bin position of bin k depends on the bin widths b'_k of all the previous bins, it is referred to as the integral nonlinearity (INL_k),

$$DNL_k = b'_k - b$$
$$INL_k = t'_k - t_k \,.$$

(2.21)

If only one value for the differential nonlinearity or for the integral nonlinearity is given, this refers to $DNL = \max|DNL_k|$ or $INL = \max|INL_k|$.

Besides affecting the bin position, non-idealities of the binning naturally also affect the expected number of photons that fall into a bin. The optical power $P_{opt}(t)$ that is incident on the detector corresponds to a photon rate $p(t)$ times Planck's constant h times the optical frequency f_0,

$$P_{opt}(t) = p(t)hf_0 \,.$$

(2.22)

The expected number of photons z_k that is detected within bin k of bin width b'_k and which bin is centered at time t'_k is given by

$$z_k = \int_{t'_k - b'_k/2}^{t'_k + b'_k/2} p(t_1)dt_1 \,.$$

(2.23)

For the case of constant optical power $P_{opt}(t) = const$, and thus $p(t) = const$, the expected number of photons z_k in each of the k bins of width b'_k simplifies to

$$z_k = b'_k \, p \,.$$

(2.24)

As shown in Fig. 2–14 (c), for constant optical power the expected number of photons is identical for all bins of the ideal binning. However, for the non-ideal binning in Fig. 2–14 (d) the expected number of photons represents the non-ideal bin width.

Binning non-idealities impair the detection of intensity modulated signals, because the detection of a higher optical power cannot be distinguished from a larger bin width.

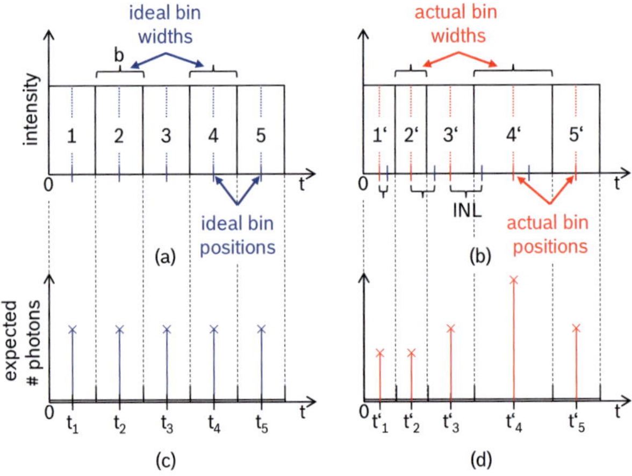

Fig. 2–14: Binning non-idealities. (a) Idealized case with time slots (bins) of identical widths $b_k=b$ that are equidistantly spaced in time at positions $t_1...t_5$. (b) Non-ideal, real-world implementation with unequal bin widths $b'_1...b'_5$ and non-equidistant spacing in time at positions $t'_1...t'_5$. Expected number of photons per bin (c) for an ideal binning and (d) for a non-ideal binning. The difference between actual bin width and ideal bin widths is referred to as differential nonlinearity (DNL). The time difference between actual bin position (red) and ideal bin position (blue) is referred to as integral nonlinearity (INL).

Niclass et al. [9] implemented a 3D camera based on time-correlated single-photon counting (TCSPC) which reaches 97.6 ps timing resolution and covers 100 ns time of flight corresponding to a maximum range of 15 m. The ranging experiments up to 3.6 m target distance result in a systematic distance error $\varepsilon_\mu = 9$ mm and a repeatability error $\varepsilon_\sigma = 5.2$ mm. With the error definition in Appendix A, this corresponds to a measurement accuracy of $\varepsilon(3.6\,\text{m}) = 19.4$ mm for a target distance up to 3.6 m. Over the entire 15 m measurement range, the measured maximum INL is 180 ps. Using $\Delta\tau = \text{INL}$ in Eq. (2.3), the INL translates into a systematic distance error $\varepsilon_\mu = 27$ mm for a target distance up to 15 m. This measurement accuracy is not sufficient for the target application here.

Richardson [35] implemented a timing circuit based on a gated ring oscillator, which reaches a resolution of 52 ps at a DNL of ~21 ps and maximum INL of ~73 ps. The maximum INL corresponds to a systematic distance error $\varepsilon_\mu = 11$ mm.

It should be noted that commercially available real-time oscilloscopes [36] reach sampling rates of 120 GHz with bin widths $b \leq 10\,\text{ps}$. Their circuits are implemented in dedicated silicon-germanium (SiGe) RF processes. The timing circuits by Niclass and Richardson mentioned above are compatible with SPAD co-integration in conventional CMOS processes and do not require dedicated SiGe processes. Further, optimization for imaging poses some additional constraints on the RF capability of the semiconductor process. For good photon detection efficiency the thickness and the separation of the layers of the CMOS process are reduced. A thin layer thickness and layer separation reduces photon absorption in the layers that are not used for photo-detection. However, a thin layer separation also increases a parasitic capacitance. Analogous to a parallel-plate capacitor, the parasitic capacitance between transmission lines in different layers increases with decreasing layer separation. This increased capacitance impairs the RF performance. Moreover, certain process layers that could improve the RF performance might not be available at all in the photosensitive area [37]. Therefore a ranging method with limited bandwidth requirements such as frequency-domain reflectometry would be advantageous in combination with SPADs.

2.3.3.2 Frequency-Domain Reflectometry with SPADs

Single-photon synchronous detection (SPSD) [38] is an adaption of time-of-flight distance measurement in the frequency domain which adaptation is suitable for SPADs. The phase difference between sent and received signal is evaluated. The concept is similar to modulated pixels (see 2.3.2, Fig. 2–12) since the light source is modulated in synchronism with the receiver. Fig. 2–15 shows the core components of an SPSD system.

An RF driver generates a modulation frequency $f_{\mathrm{m}} = 1/T_{\mathrm{m}}$. The RF driver continuously modulates the current of a laser diode, which sends out a sinusoidal intensity modulated laser beam to the target (sent signal). Some of the backscattered light (received signal) is collected with a receiver lens and detected with a single-photon avalanche diode (SPAD). The output of the SPAD is a time series of output pulses of constant height and width. Each pulse represents at least one photon at one instant of time. The pulse rate over time is proportional to the received light intensity I_{Rx} over time. It should be noted that the SPAD itself is not modulated. The SPAD pulses are fed to a 'binning' component for sampling the received signal. The binning process can be thought of as a 'time de-multiplexer' that is operated in synchronism with the RF modulation frequency from the RF driver. The outputs of the de-binning are connected to counters C0…C3.

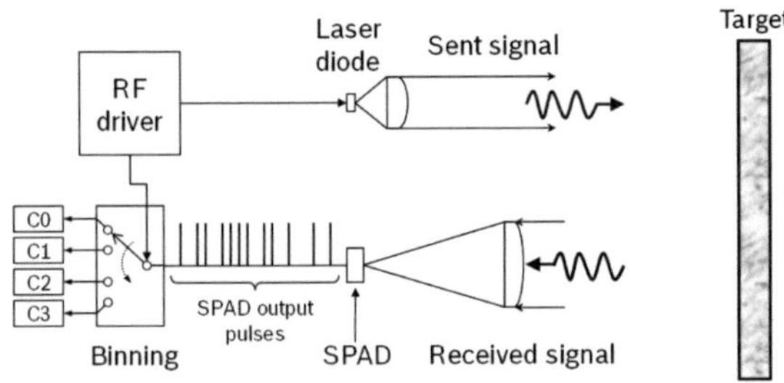

Fig. 2–15: Single-photon synchronous detection (SPSD) system for frequency-domain reflectometry with a single-photon avalanche diode (SPAD). An RF driver continuously modulates the current of a laser diode, which sends out a sinusoidal intensity modulated laser beam to the target (sent signal). The backscattered light (received signal) is collected with a receiver lens and detected with a SPAD. The output of the SPAD is a time series of output pulses that are fed to the binning. The binning process can be thought of as a 'time de-multiplexer' that is operated in synchronism with the RF modulation frequency from the RF driver. The outputs of the de-binning are connected to counters C0…C3.

The binning process is described in more detail with reference to Fig. 2–16. The RF driver has two outputs. The first output modulates the current of the laser diode. An inset shows one period of the sinusoidal modulation of the current $i_{m,Tx}$ over time. Note the inverse time axis - the signal at time $t = 0$ is first at the laser diode. The second output of the RF driver operates the time de-multiplexer of the binning in synchronism with the laser diode modulation. Pulses incident during the first quarter of the RF-period are directed to a first counter C0, corresponding to bin 0, pulses incident during the second quarter of the RF-period into a second counter C1, corresponding to bin 1, and so on. The counter values $x_0…x_3$ are depicted on the right as a histogram over time. The histogram is built up over multiple modulation periods. The histogram covers the duration of one modulation period T_m, since the output signal and the binning repeat every modulation period. Analogous to the modulated pixels in Section 2.3.2, the phase of the envelope of the received signal intensity at the fundamental modulation frequency can be calculated from the counter values using Eq. (2.19). The distance is again calculated from Eq. (2.9).

In contrast to modulating the detector in previous implementations of frequency-domain reflectometry, the SPAD detector is not modulated in SPSD. The synchronous mixing of RF signal from the RF driver and the received signal intensity is performed by the binning process. Recall that the modulated pixel in [32] operates at a modulation frequency of 20 MHz. The binning process for SPSD can be implemented as a time de-multiplexer in the digital domain at high frequencies of several hundred MHz or even some GHz. The impact of the modulation frequency on distance accuracy will be shown in more detail in the system performance simulation in Section 5.4.2 (Fig. 5–46).

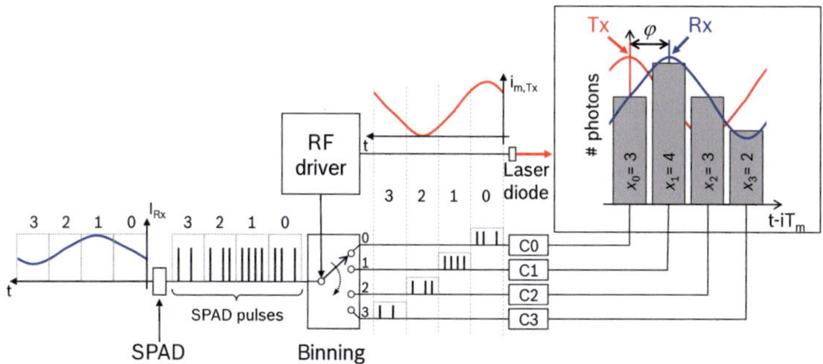

Fig. 2–16: Binning process. The RF driver has two outputs. The first output modulates the current of a laser diode. The inset shows the sinusoidal modulation of the current $i_{m,Tx}$ over time. The second output of the RF driver operates the time de-multiplexer of the binning in synchronism with the laser diode modulation. Pulses incident during the first quarter of the RF-period are directed to a first counter C0, corresponding to bin 0, pulses incident during the second quarter of the RF-period into a second counter C1, corresponding to bin 1, and so on. The counter values $x_0 \ldots x_3$ are depicted on the right as a histogram over time. One modulation period T_m is shown since the signal and binning repeat periodically. The phase of the received signal (Rx) is calculated from the counter values and compared to the phase of the transmitted signal (Tx). The target distance is calculated from the phase difference φ between sent and received signal.

A first successful implementation of single-photon synchronous detection (SPSD) has been published by Niclass et al. in [9, 38, 39]. Their research group at the École Polytechnique Fédérale de Lausanne (EPFL) built up a prototype of a SPSD 3D camera with a systematic distance error $\varepsilon_\mu = 110\,\text{mm}$ and a repeatability error $\varepsilon_\sigma = 38\,\text{mm}$ at a maximum target distance of $L = 2.4\,\text{m}$. With the error definition in Appendix A, this corresponds to a measurement accuracy $\varepsilon(2.4\,\text{m}) = 186\,\text{mm}$ for a target distance up to 2.4 m. Even though the performance is by far not sufficient for a millimeter-precision laser rangefinder, this work is a successful demonstration of the SPSD principle.

Especially the binning procedure leaves room for further improvement. The binning in [38, 39] has two bins per modulation period $T_m = 1/f_m$. Each bin covers 50 % of the modulation period with a bin width $b = T_m/2 = 1/2f_m$. A bin can be seen as a rectangular sampling window with a finite width b. As shown in Appendix C, the sampling with a finite sampling window in time domain attenuates the signal with a sinc transfer function in frequency domain. In other words, the bin of finite width acts as a filter on the received signal. The received signal at the modulation frequency f_m is attenuated significantly to $H(f_m) = \mathrm{sinc}(bf_m) = \mathrm{sinc}(\pi/2) \approx 63.7\%$. The attenuation at the modulation frequency can be reduced by using narrower sampling windows. However, no signal energy should be lost. Therefore still the entire modulation period should be covered, however, with an increased number of bins. For four (eight) bins per period, 90.0 % (97.4 %) of the signal at the modulation frequency are recovered.

A portion of the systematic distance error in [39] is explained by signal distortion due to higher harmonics in the intensity modulated laser beam. This distortion can be mitigated by sending a clean sinusoidally intensity modulated signal or by oversampling in order to reduce aliasing.

The system in [38, 39] is a 3D camera for capturing distances at video frame rates. The acquisition time per frame is 45 ms. For a laser rangefinder, a longer integration time of up to 4 s can be used, which will reduce the repeatability error due to noise.

Furthermore, the system runs at a relatively low modulation frequency of 30 MHz. Since, according to Eq. (2.10), the distance error ΔL scales with the inverse of the modulation frequency higher, but not yet critical, modulation frequencies of several 100 MHz to 1 GHz are therefore beneficial if the phase accuracy is limited. However, high modulation frequencies already give rise to new problems, such as unequal sampling window widths (bin widths) that will be addressed and mitigated in this thesis.

In conclusion, the potential of the SPSD approach has not fully been leveraged yet. As mentioned previously, frequency-domain reflectometry is beneficial for a low-cost laser rangefinder. A narrow-band signal can be used with the respective benefit of low-cost electronics with limited bandwidth. SPSD enables the combination of frequency-domain reflectometry with CMOS-compatible SPADs. Therefore, SPSD is a promising approach for a simple, yet accurate low-cost system that will be further investigated in this thesis.

	Bosch GLM 250 laser rangefinder [4]	PMDtec CamCube 3.0 3D camera [27]	Swiss Ranger SR-4000 3D camera [26]	EPFL[*] TCSPC 3D camera [9]	EPFL[*] SPSD 3D camera [39]	Target for this work
Technology	Gain modulated APD	Modulated CCD/CMOS pixel	Modulated CCD	CMOS SPAD	CMOS SPAD	CMOS SPAD
Range L	0.05 ... 250 m	0.3...7.0 m	5.0 m (10.0 m)	15 m	5.0 m	20 m
Range dependent accuracy	$\pm$ (1.5 mm + 0.05 mm/m $\times L$)	< 3 mm (1σ, 4m distance, 75 % target reflectivity)	4 mm (6 mm), (1σ, 2 m distance); $\pm$ 10 mm ($\pm$ 15 mm) (meas. range, 99% target reflectivity)	$\varepsilon_\mu = 9$ mm, $\varepsilon_\sigma = 5.2$ mm (3.6 m distance)	$\varepsilon_\mu = 110$ mm, $\varepsilon_\sigma = 38$ mm (2.4 m distance)	$\pm$ (2.0 mm $\pm$ 0.1 mm/m $\times L$)
Temperature drift of accuracy	<0.5 mm	120 mm warm-up over 40 min [40]	$\leq$1.5mm/°C (10-50°C)	n.a.	n.a	n.a.
Modulation frequency	1 GHz	20 MHz	30 MHz (15 MHz)	-	30 MHz	400MHz – 1 GHz
Measurement time	typ. < 0.5 s, max. 4 s	12.5-25 ms (40-80 fps)	$\geq$ 20 ms ($\leq$50 fps)	50 ms	45 ms	max. 4 s
Illumination	635 nm laser, class II, eye safe	870 nm	850 nm	850 nm	850 nm	635 nm laser, class II, eye safe
Background light tolerance	100,000 lx	n.a.	n.a.	n.a. (measured at 150 lx)	n.a.	1,000-10,000 lx
Housing	IP54, drop test on concrete floor	Commercial reference design	Industrial grade	Laboratory prototype	Laboratory prototype	Laboratory prototype

Table 2.1: Performance data of commercial laser rangefinder, state-of-the-art integrated ranging systems and target system;

*EPFL = École Polytechnique Fédérale de Lausanne;

2.4 Conclusion

The target application of this work, measuring distances on a construction site with millimeter precision, requires the use of an active distance measurement technique that also works on featureless, extended, plain surfaces. Distance measurement by reflectometry is preferred for a compact, handheld laser rangefinder. Frequency-domain reflectometry further allows the use of low-cost components with limited bandwidth. Preferentially, the phase of the envelope of a continuously intensity modulated laser is evaluated for distance measurement.

Integrated ranging systems, have the potential to outperform state-of-the-art semi-integrated laser rangefinders in terms of cost, size and manufacturability. Single-photon avalanche diodes (SPADs) are attractive photodetectors for an integrated ranging system because they are compatible with existing semiconductor fabrication processes. The digital output pulses of a SPAD can directly be used for digital signal processing.

Single-photon synchronous detection (SPSD) combines the benefits of distance measurement by frequency-domain reflectometry with the benefits of SPADs. Hence, SPSD is a promising approach for a simple, yet accurate low-cost laser rangefinder and will be further investigated in this thesis.

3 CMOS Single-Photon Avalanche Diodes (SPADs)

One focus of this thesis is to understand how SPAD properties affect a ranging system. The SPAD component itself is not designed as a part of this work. The research of Rochas [41], Niclass [9] and Richardson [35] provides further information about SPAD design, characterization and fabrication.

This chapter starts with a brief review of the fundamental physics of conventional photodetectors (Section 3.1). The properties and peculiarities of single-photon avalanche diodes (SPADs) are explained (Section 3.2). Section 3.3 focuses on SPAD technology and different devices leading to the CMOS SPADs used in this research. Section 3.4 provides further details about the mandatory external circuitry for operating a SPAD. Since the target of this research is a laser rangefinder that measures a distance based on time of flight of a signal from the rangefinder to the target and back to the rangefinder, the timing of a SPAD requires special attention and is highlighted separately in Section 3.5. A conclusion regarding SPADs is provided in Section 4.3 after the experimental characterization of SPADs in Chapter 4.

3.1 Fundamental Physics of Photodetectors

This Section briefly reviews the fundamental physics of the conventional photodetectors pin-diode and avalanche photodiode (APD).

3.1.1 PIN-Diode

A pin-diode for photodetection is a reverse-biased pn-junction. Photons with energies $W_{photon} = hf_0 > W_G$ (optical frequency f_0, Planck's constant h) larger than the bandgap energy W_G that are incident on the pn-junction generate, with a certain probability, electron-hole pairs (e-h pairs). The probability that an e-h pair is generated per absorbed photon is the quantum efficiency $\eta_{quantum}$. In the electric field of the reverse biased pn-junction, the electrons drift towards the n-contact and the holes towards the p-contact. For one e-h pair one elementary charge e is transferred in an external circuit. This generates a photocurrent i proportional to the optical power P that is incident on the detector,

$$i(t) = \frac{\eta_{\text{quantum}} \, e}{hf_0} P(t) \,. \tag{3.1}$$

The proportionality constant between the photocurrent i and the optical power P is referred to as the responsivity or sensitivity S of the detector. The sensitivity defines the photocurrent per incident optical power in [A/W],

$$S = \frac{\eta e}{hf_0} \,. \tag{3.2}$$

The optical power P is proportional to the optical intensity or irradiance I in [W/m^2] and the active area of the photodetector A_{det} (assuming that the entire active area is illuminated),

$$P(t) = A_{\text{det}} \, I(t) \,. \tag{3.3}$$

One can simply increase the photocurrent, Eq.(3.1), by increasing the active area, Eq. (3.3).

On a physical level, the optical power P can be expressed by the rate of incident photons p and the photon energy $W_{photon} = hf_0$. Thus the photon rate p is given by

$$p(t) = \frac{P(t)}{W_{\text{photon}}} = \frac{P(t)}{hf_0} \,. \tag{3.4}$$

With Eq. (3.4), Eq. (3.1) can be rewritten in terms of the electron rate i/e and the rate of actually absorbed photons $\eta_{\text{quantum}} \, p$,

$$\frac{i(t)}{e} = \eta_{\text{quantum}} \, p(t) = \eta_{\text{quantum}} \, \frac{P(t)}{hf_0} \,. \tag{3.5}$$

Fig. 3–1 shows the structure of a pin-diode together with the electric field distribution inside under reverse-bias operation [42]. An intrinsic layer between the p-doped and the n-doped semiconductor regions increases the width of the depletion or space-charge region. Electrons (holes) generated in the depletion region are accelerated by the electric field and drift towards the n-contact (p-contact).

Furthermore, carriers that are generated outside of the depletion region can diffuse to the depletion region and contribute to the photocurrent. However, diffusion is an inherently slow process. The mean diffusion velocity is about

1μm/ns or less [42]. Therefore, diffusion impairs the bandwidth of photodetectors. For a fast detector, most of the light should be absorbed in the depletion region. This is achieved by decreasing the widths of the p-region (d_p) and n-region (d_n) and by increasing the width (w_A) of the i-region (absorption region) of the pin-photodiode. A high doping level is used for the shortened p- and n-region.

Fig. 3–1 also illustrates the absorption of light in the semiconductor [42]. The light power P_e, incident externally from the left, is partially reflected with a power reflection factor R_P. The power inside the semiconductor $P_i(x)$ is attenuated exponentially. The absorption coefficient α corresponds to the reciprocal absorption length $1/\alpha$.

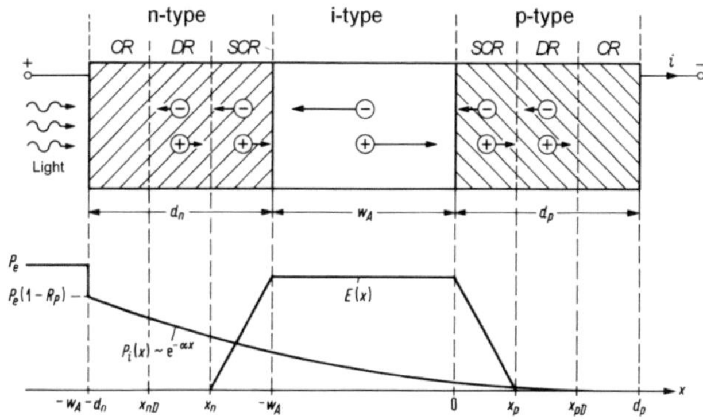

Fig. 3–1: Schematic of a pin-diode (not drawn to scale). CR contact region, DR diffusion region, SCR space-charge (or depletion) region. P_e light power incident from region external of the semiconductor, R_P power reflection factor of the semiconductor surface, $P_i(x)$ light power inside the semiconductor, α absorption coefficient, d_n (d_p) length of n-doped (p-doped) semiconductor, w_A length of intrinsic absorption zone, $E(x)$ x-component of electric field (modified from [42] © Grau, Freude 1991).

If the irradiated n-region is much shorter than the absorption length, i.e., $\alpha d_n \to 0$ in Fig. 3–1, the light power in the i-region reads

$$P_i(x,t) = P_e(t)(1-R_P)e^{-\alpha(x+w_A)} \ . \tag{3.6}$$

The power fraction which is absorbed inside the i-zone represents the quantum efficiency η_{quantum} of the photodiode,

$$\eta_{\text{quantum}} = \frac{P_i(-w_A,t) - P_i(0,t)}{P_e(t)} = (1 - R_P)(1 - e^{-\alpha w_A}). \tag{3.7}$$

Based on Eq. (3.7), the quantum efficiency can be influenced by the power reflection factor R_P, the absorption coefficient α and the absorption length w_A.

For an air-silicon interface (refractive indices $n_{\text{air}} = 1$, $n_{\text{silicon}} = 3.9$ at an optical wavelength $\lambda_0 = 635\,\text{nm}$ [43]), the power reflection factor of the interface is

$$R_P = \left(\frac{n_{\text{silicon}} - n_{\text{air}}}{n_{\text{silicon}} + n_{\text{air}}} \right)^2 \approx 35\%. \tag{3.8}$$

In high and medium price detectors, the detector surface is equipped with an anti-reflection coating to reduce the surface reflectivity. For the targeted low-cost system, however, it is more economical to increase the diameter of the plastic receiver lens instead of applying an expensive coating. For the example above, the diameter of the lens must be increased by 24 % in order to compensate for the surface reflectivity of the air-silicon interface.

The absorption coefficient α is material specific and wavelength dependent. Although SPADs have been demonstrated in various semiconductor materials (e.g. Si [44], InGaAs-InP [45], SiGe [46]), cost and co-integration requirements in standard fabrication processes dictate the use of silicon. At $\lambda_0 = 635\,\text{nm}$ wavelength the absorption length in silicon is $1/\alpha \approx 3.1\,\mu\text{m}$ [42]. The absorption coefficient of silicon decreases at higher wavelengths. A poor sensitivity in the near infrared is therefore expected. At wavelengths $\lambda_0 > 1.1\,\mu\text{m}$, photons are no longer absorbed because of the 1.1eV bandgap of silicon $W_{BG} = hf_0 = hc/\lambda_0$, where h is Planck's constant and c the vacuum speed of light.

The width w_A of the absorption zone can essentially be chosen by design. To achieve 50 % efficiency, an absorptions zone width of $w_A = -\ln(1 - \eta_{\text{quantum}})/\alpha \approx 2.2\,\mu\text{m}$ is required (using Eq. (3.7) and assuming $R_P = 0$). There is a trade-off between quantum efficiency and high-speed operation. For high quantum efficiency, w_A should be as large as possible. However, high-speed operation requires small carrier transit times in the depletion region, i.e., small w_A. It should be noted that there is also an optimum width for high-speed operation, since the depletion-layer capacitance increases with decreasing w_A [42].

In practice there are various layers on top of the actual active region where photons are absorbed but do not reach the depletion region. These layers decrease the quantum efficiency of the detector.

3.1.2 Avalanche Photodiode

Much like the pin-diode, an avalanche photodiode (APD) is a reverse-biased pn-junction. Fig. 3–2 shows the structure of an APD together with the electric field distribution inside it under reverse-bias operation [42].

A photon happens to generate a primary e-h pair with a probability η_{quantum}. In case that $w_L + d_p << w_A$ (or if the avalanche zone and the p+-region have a bandgap $W_{BG} > hf$), the light incident from the right side of the APD is absorbed in the absorption zone. Due to the electric field, one type of carriers, in this case the holes, drifts to the avalanche or multiplication zone. Due to a high reverse bias voltage, but yet below breakdown, charges within the multiplication zone of the junction are accelerated and create secondary e-h pairs by impact ionization.

Without avalanche multiplication, for one primary e-h pair one elementary charge e is transferred to an external circuit. The photocurrent from primary carriers i_p is given by Eq. (3.1) with $i = i_p$. With avalanche multiplication, for one primary e-h pair the charge eM_0 will on average be transferred to an external circuit, where M_0 is the average multiplication factor or avalanche gain.

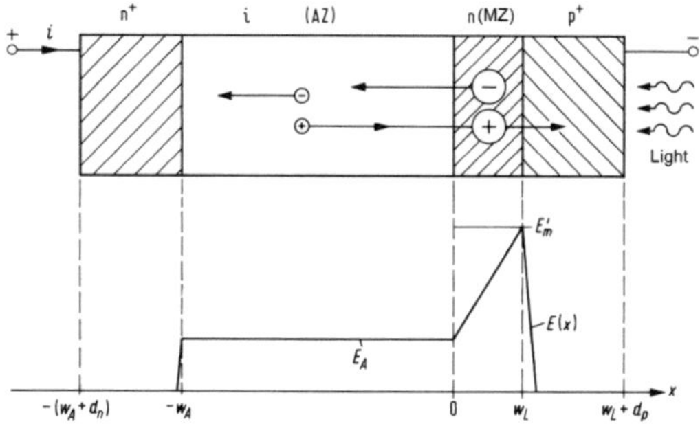

Fig. 3–2: Schematic of an avalanche photodiode (APD) (not drawn to scale). AZ absorption zone (length w_A), MZ multiplication zone (avalanche zone) (length w_L), $E(x)$ x-component of electric field, E_A electric field in absorption zone, E_m maximum electric field strength (modified from [42] © Grau, Freude 1991).

The photocurrent i_A of the APD is proportional to the avalanche gain M_0 and the photocurrent i_p from primary e-h pairs,

$$i_A(t) = M_0 i_p(t).$$
(3.9)

Analogous to Eq. (3.5), Eq. (3.9) can be rewritten in terms of the electron rate i/e and the rate of actually absorbed photons $\eta_{\text{quantum}} p$ times the avalanche gain M_0,

$$\frac{i(t)}{e} = M_0 \eta_{\text{quantum}} p(t).$$
(3.10)

The ionization coefficients α_i (β_i) characterize the capability of electrons (holes) to cause impact ionization; $\alpha_i \, dx$ ($\beta_i \, dx$) is the expected number of e-h pairs that an electron (hole) generates when passing a length dx [42]. The ionization coefficients depend on the material and the strength of the electric field. In silicon, electrons dominate the ionization process.

For the case $\beta_i / \alpha_i = 0$ only electrons generate secondary e-h pairs by impact ionization, whereas holes do not. Once all of the primary and secondary electrons have left the multiplication zone, no further e-h pairs are generated. After any remaining holes have propagated through the device, the avalanche current eventually stops by itself. The avalanche process is self-limiting. A typical multiplication factor M_0 of an APD is around 100 at 100 V bias voltage in silicon [42].

For the case $\beta_i / \alpha_i \neq 0$ there is a positive feedback mechanism because not only electrons but also holes can create secondary e-h pairs. For the same α_i as above, this positive feedback increases the multiplication factor M_0. The ratio β_i / α_i depends on the strength of the electric field. At a sufficient field strength the ratio approaches 1, $\beta_i / \alpha_i \to 1$, and the positive feedback eventually leads to an infinite avalanche gain $M_0 \to \infty$. The avalanche becomes self-sustaining, i.e., the avalanche keeps burning once that it has been started. The voltage at which the avalanche becomes self-sustaining is the breakdown voltage V_{bd}.

3.2 Properties and Peculiarities of SPADs

This section introduces single-photon avalanche diodes (SPADs) and explains the performance parameters that are used in course of this work.

3.2.1 What is a SPAD?

From a system point of view, a SPAD is effectively a triggered photonic device that generates a digital output pulse when struck by a single photon.

A SPAD is a reverse biased pn-junction that can be temporarily biased beyond the steady-state breakdown voltage V_{bd} [44]. The behavior of the device in this operating mode differs considerably from a conventional avalanche-photodiode (APD). At first glance, biasing a pn-junction above breakdown seems inappropriate.

Fig. 3–3 shows the current-voltage curve of a pn-junction with bias conditions for a pin-diode, an APD and a SPAD [47]. Space-time diagrams, as introduced by Aull et al. [48], are shown as insets in Fig. 3–3 (a)-(c). These diagrams offer a neat visualization of carrier multiplication in the space-charge region of the pn-junction for the different modes of operation.

Photons incident on the pn-junction generate, with a certain probability $\eta_{quantum}$, electron-hole pairs (e-h pairs). For the example of a <u>pin-diode</u>, Fig. 3–3 inset (a), a photon happens to generate a primary e-h pair. For one e-h pair one elementary charge e is transferred in an external circuit. In <u>APDs</u>, Fig. 3–3 inset (b), additional secondary e-h pairs are generated by impact ionization. In the chosen example drawing, only electrons generate secondary e-h pairs by impact ionization, whereas holes do not ($\beta_i / \alpha_i = 0$). The avalanche process is self-limiting. For one e-h pair the charge eM_0 will on average be transferred in an external circuit, where M_0 is the avalanche gain.

<u>SPADs</u> are pn-junctions temporarily operated beyond breakdown voltage. Any free charge carrier, electron or hole, within the depletion region can start a strong, self-sustaining avalanche. Assuming that there are no free carriers in the depletion region, at least for a limited amount of time, the device can be biased above breakdown voltage without starting an avalanche. Recent advances in semiconductor technology allow time delays up to tens of milliseconds before a non-photon-induced breakdown occurs [33]. The first carrier, either photo-generated, thermally generated or tunneling into the junction, however, starts the avalanche. Both the electrons and the holes generate secondary e-h pairs, as

sketched in Fig. 3–3 inset (c). Since the device is operated above breakdown voltage, the avalanche gain is virtually infinite and the avalanche is self-sustaining. The corresponding strong avalanche current can be detected. No external amplifiers are needed. Hence, the SPAD is capable of detecting a single photon.

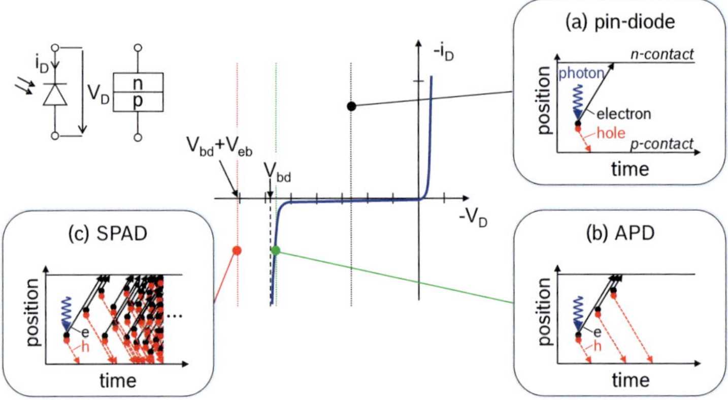

Fig. 3–3: Current-voltage-curve of a pn-junction with bias conditions for a pin-diode, an APD and a single-photon avalanche diode (SPAD). i_D diode current, V_D voltage over the diode. The SPAD is biased above breakdown voltage V_{bd} with an excess bias voltage V_{eb}. The insets (a)-(c) show carrier multiplication in the space-charge region of the pn-junction for the different modes of operation.
(a) pin-diode: A photon (blue) generates one electron-hole pair (e-h pair). In the electric field of the reverse-biased pn-junction, the electron (black) drifts towards the n-contact and the hole (red) towards the p-contact.
(b) APD: A photon generates a primary e-h pair. The electron, but not the hole, starts an avalanche of additional secondary e-h pairs by impact ionization. The number of carriers increases with a multiplication factor M_0. The current stops by itself, when the carriers have passed through the device. The avalanche is self-limiting.
(c) SPAD: A photon generates a primary e-h pair. Both the electrons and the holes can cause additional secondary e-h pairs by impact ionization. Hence, both the number of electrons and the number of holes multiply and result in a self-sustaining avalanche, i.e., in an avalanche that keeps burning once that it has been started.

However, this self-sustaining avalanche must be stopped to prevent damaging the device and to be able to detect another photon. For stopping the avalanche, the voltage across the SPAD is lowered below the breakdown voltage. The process of stopping the avalanche is called quenching. In order to detect another photon, the SPAD has to be biased above breakdown again. After restoring bias conditions ('recharge'), the SPAD is rearmed and ready to detect another photon. This triangle of triggering the avalanche (1), quenching (2) and recharging the device (3) is shown in Fig. 3–4. So-called 'front-end circuits' to perform this function are mandatory and will be explained more detailed in Section 3.4. The output signal of a single SPAD with associated front-end circuitry is an output pulse with constant width and constant height. The photocurrent or the number of electrons per detected photon is not determined by a multiplication factor or the responsivity, but merely depends on the external circuitry that determines the SPAD output pulse.

It should be noted that the SPAD output pulse looks essentially the same regardless of whether the avalanche has been triggered by one single photon or by more than one photon, since the pulse shape is determined by the front-end circuit.

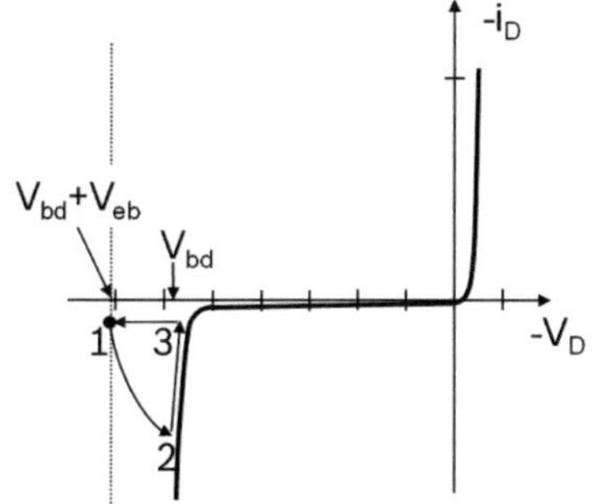

Fig. 3–4: SPAD operation triangle. A photon triggers the avalanche and starts a self-sustaining avalanche (1). External circuitry stops the avalanche by decreasing the voltage below the breakdown voltage V_{bd} (2). The SPAD is recharged to the excess bias voltage V_{eb} above breakdown (3) to be able to detect another photon.

The SPAD cycles the triangle shown in Fig. 3–4 for each detected photon and generates an output pulse. In other words, an output pulse or SPAD count is an avalanche event that is recognized by the subsequent digital circuitry. For a series of detected photons, the SPAD output is a time series of pulses of constant height and width as sketched in Fig. 3–5. Therefore, SPADs are also referred to as Geiger mode APDs or G-APDs. The name originates from Geiger-Müller counters for measuring ionizing radiation.

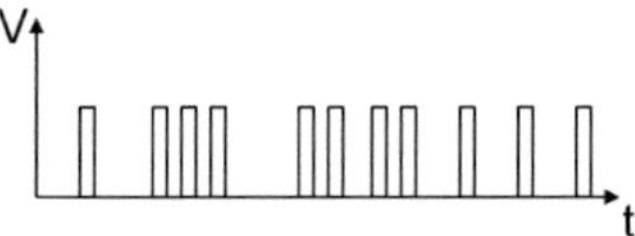

Fig. 3–5: SPAD output. The output of the SPAD is a series of pulses of constant height and width. The count rate is proportional to the photon rate.

3.2.2 Photon Statistics

In a conventional photodetector, the detector output represents an average intensity and thus an average number of photons. A SPAD, however, is capable of detecting single photons. The output of a SPAD is a train of pulses that represent individual incident photons (at least those that manage to start a self-sustaining avalanche which is recognized by the subsequent digital circuitry). Thus the photon statistics of the photons incident on the SPAD will also be reflected by the output pulses of the SPAD. The following brief description is based on [42]. A detailed analysis of different light sources with their respective characteristics and further references can be found in [49].

Bose-Einstein distribution

Photons from a thermal light source, for example the sun or an incandescent light bulb, follow a Bose-Einstein distribution [50]. The probability of observing N photons in the z modes of the thermal light source and the corresponding variance are given by

$$\text{Prob}(N) = \frac{(N+z-1)!}{N!(z-1)!} \frac{1}{\left(1+\dfrac{N}{z}\right)^z + \left(1+\dfrac{z}{N}\right)^N}$$

$$\sigma_N^2 = \overline{(N-\overline{N})^2} = \frac{\overline{N}^2}{z} + \overline{N} \tag{3.11}$$

For a thermal light source, the number of modes z is much larger than the average number of photons $\overline{N} \ll z$.

Poisson distribution

For $\overline{N} \ll z$, $N \ll z$ the Bose-Einstein distribution in Eq. (3.11) converges to the Poisson distribution,

$$\mathrm{Prob}(N) = \frac{\overline{N}^{N}}{N!}\exp(-N) \;,\quad \sigma_N^2 = \overline{N}\,.\tag{3.12}$$

The Poisson distribution gives the distribution of statistically independent classical events. The average number of photons $\overline{N}$ and the variance σ_N^2 are equal. For the case that few photons N are distributed over a large number of possible modes z, the probability of having several photons in the same mode is negligibly small. Thus, with a high probability, the photons of a thermal light source will behave as if they were classical, independent events [42].

The Poisson distribution also applies for laser light [51, 52]. However, in contrast to a thermal light source, the Poisson distribution also applies for a large number of photons per mode, $\overline{N} \gg z$.

The variance σ_N^2 and thus the standard deviation σ_N serves as a measure of the uncertainty of the detected number of photons. In the context of this thesis, the standard deviation is used as a measure for the photon noise. Since $\sigma_N = \sqrt{N}$, the absolute photon noise increases with an increasing number of photons. However, the relative photon noise defined as

$$\frac{\sigma_N}{\overline{N}} = \frac{\sqrt{\overline{N}}}{\overline{N}} = \frac{1}{\sqrt{\overline{N}}}\tag{3.13}$$

decreases with an increasing number of photons. Thus, the impact of photon noise scales with $1/\sqrt{N}$ and will reduce for a higher photon rate or a longer integration time.

Inter-arrival time

The inter-arrival time Δt is the time between two subsequent (detected) photons. As shown in Fig. 3–5 the output of a SPAD is a series of pulses that represent detected photons. The inter-arrival time is thus measured by the time difference between subsequent pulses.

For Poisson distributed photons and a constant photon rate p, the probability density distribution of the inter-arrival time Δt between two subsequent photons is given by an exponential decay (see derivation in Appendix B),

$$f(\Delta t) = p\exp(-p\Delta t). \tag{3.14}$$

This probability density distribution is shown for two different photon rates p_1 and $p_2 > p_1$ in Fig. 3–6.

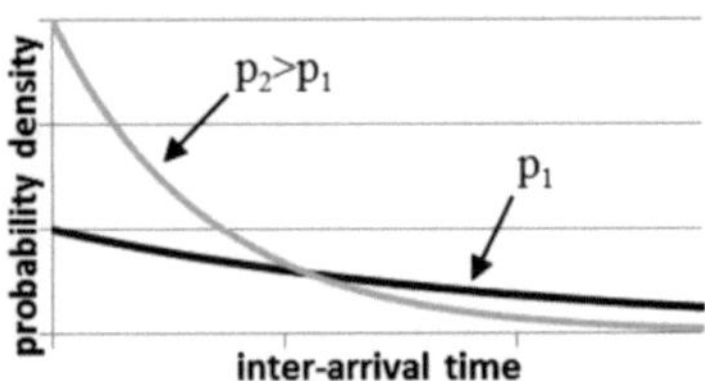

Fig. 3–6: Inter-arrival time plot for two photon rates p_1 and $p_2>p_1$.

A plot of the probability density distribution of the inter-arrival time will later be used to analyze the measured pulse train of the SPAD output. The expected distribution due to Poisson distributed photons is an exponential decay as shown in Fig. 3–6. Thus deviations of a measured inter-arrival time distribution from the expected distribution can provide information about a non-ideal behavior of the SPAD.

3.2.3 Performance Parameters

SPADs can be characterized by a set of performance parameters. The parameters listed here are described in more detail in the sections below.

- Count Rate and Photon Detection Efficiency — (Section 3.2.3.1)
- Dark Count Rate — (Section 3.2.3.2)
- Dead Time — (Section 3.2.3.3)
- Detector Noise and Dynamic Range — (Section 0)
- Afterpulsing — (Section 3.2.3.5)
- Jitter — (Section 3.2.3.6)
- Supply Voltages — (Section 3.2.3.7)
- Array Fill Factor — (Section 3.2.3.8)
- Crosstalk — (Section 3.2.3.9)

The tradeoffs for the performance parameters are discussed later in this section. A SPAD can be engineered for a specific application.

3.2.3.1 Count Rate and Photon Detection Efficiency

The count rate n is given by the number of output pulses N during a measurement time t_{meas},

$$n(t) = \frac{N}{t_{\text{meas}}}.$$

(3.15)

The count rate is proportional to the incident photon rate p and the photon detection efficiency (PDE) η_{PDE},

$$n(t) = \eta_{\text{PDE}} \, p(t) = \eta_{\text{PDE}} \frac{P(t)}{hf_0}.$$

(3.16)

The photon detection efficiency η_{PDE} of a SPAD is defined by the probability that a photon generates an electron-hole pair (quantum efficiency η_{quantum}) times the probability that this electron-hole pair triggers a self-sustaining avalanche (avalanche efficiency $\eta_{\text{avalanche}}$),

$$\eta_{\text{PDE}} = \eta_{\text{quantum}} \, \eta_{\text{avalanche}}.$$

(3.17)

Since the occurrence of photon absorption within the semiconductor is a stochastic process, the efficiencies defined here are expected values. Expected values are also given for quantities such as count rates. Whenever values are measured or discussed it is assumed that the law of large numbers can be applied, i.e., the integration time for measuring is large such that the average measured value is close to the expected value.

The avalanche efficiency $\eta_{\text{avalanche}}$ increases with increasing excess bias voltage V_{eb}, since the electric field strength increases and thereby the ionization coefficients of electrons and holes [44].

It should be noted that at very high incident photon rates saturation effects set in because the SPAD needs some time to recover before the next photon can be detected. Under these conditions the count rate is not proportional to the photon rate any more. Saturation effects of a SPAD are described in Section 3.4.3. Typically, count rates are limited to some tens of MHz. A record count rate of 185 MHz has been published as part of this research [53].

3.2.3.2 Dark Count Rate

Dark counts are spurious pulses unrelated to incident photons. Dark counts can occur due to thermal generation, trap assisted generation or by tunneling from the valence band to the conduction band [54]. The transitions between valence and conduction band are sketched in Fig. 3–7. The energy levels of the valence band edge and the conduction band edge are indicated with W_V and W_C. Dark counts due to tunneling are observed at low temperatures, whereas thermally generated dark counts are the dominant source of dark counts at and above room temperature. Defects give rise to trap-assisted carrier generation and increase the dark count rate. Trap-assisted generation/recombination has first been described by Shockley and Hall in [55]. Furthermore, carriers that have been generated outside of the depletion region can diffuse into the depletion region and also cause an avalanche event. Fig. 3–7 shows the example of a thermally generated e-h pair where the hole diffuses into the depletion region. Afterpulsing will be explained further below (Section 3.2.3.5).

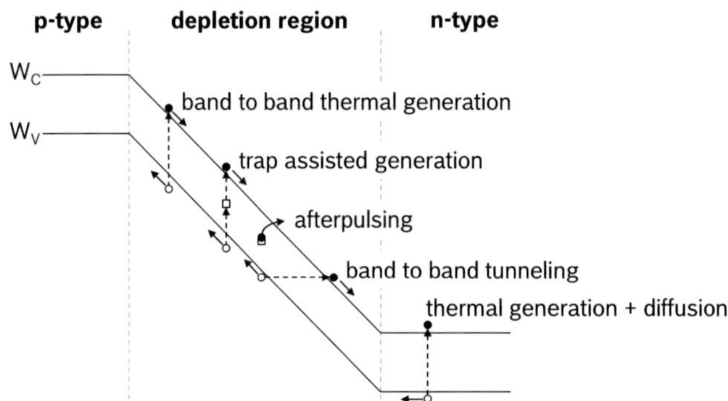

Fig. 3–7: Sources of SPAD dark counts [41]. Filled circles represent electrons, empty circles holes, squares are traps by lattice defects.

Device engineering offers some degrees of freedom to lower the dark count rate (DCR) n_{DCR}. For example, the horizontal band separation in Fig. 3–7 can be influenced with appropriate doping concentrations. A larger horizontal band separation reduces the probability of tunneling between valence and conduction band. Moreover, process steps and annealing times can be customized for low dark count rates. Details are provided in [56].

It should be noted that output pulses from dark counts are indistinguishable from photo-generated pulses. Just like a photo-generated electron-hole pair, any thermally generated or tunneling carrier can start a self-sustaining avalanche.

A larger area SPAD suffers from a higher dark count rate. Since the active area and thereby active volume of the SPAD increases, the probability of having a dark count is higher. Moreover, the probability for thermal generation increases with temperature. The DCR also increases with a higher excess bias voltage V_{eb}, since the avalanche efficiency $\eta_{\text{avalanche}}$ increases.

3.2.3.3 Dead Time

Once a photon has triggered an avalanche, the SPAD cycles trough the triangle of detection, quenching and recharging shown in Fig. 3–4. The minimum time the SPAD takes to be ready to detect another photon is the dead time (DT, t_{dead}). The dead time depends on the front-end circuitry for quenching and recharging the device. In the context of this thesis, dead time is defined as the time between detecting a first avalanche event and the moment in time when the SPAD is ready to detect a second avalanche event. During the dead time the SPAD cannot react to photons, Fig. 3–8. The SPAD is 'blind' during dead time.

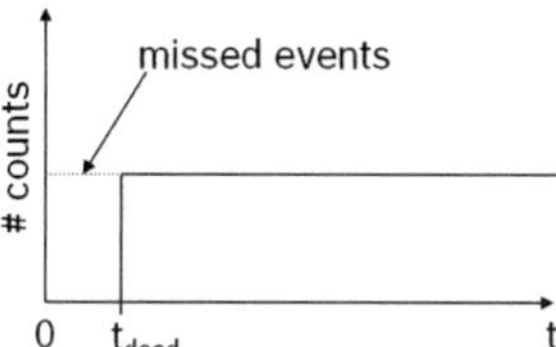

Fig. 3–8: SPAD dead time (t_{dead}). Dead time is the time after triggering a self-sustaining avalanche until the SPAD is rearmed and ready to detect another photon. Events during dead time are not counted.

Typical dead times range between several nanoseconds to some tens of nanoseconds [44, 53, 57]. The dead time poses an upper limit on the maximum count rate,

$$n_{\text{max}} = 1/t_{\text{dead}}\,.$$

(3.18)

3.2.3.4 Detector Noise and Dynamic Range

Detector noise and dynamic range are determined by the dark count rate (DCR) n_{DCR} and the maximum possible count rate n_{max}. The number of dark counts N_{DCR} in the measurement time t_{meas} is $N_{DCR} = n_{DCR} t_{meas}$.

Because there are no external amplifiers, SPAD pulses unrelated to incident photons (dark counts) are the only noise source besides photon noise. The variance of Poison-distributed dark counts with an expected value of N_{DCR} counts is $\sigma^2_{N_{DCR}} = N_{DCR} = n_{DCR} t_{meas}$. Thus the standard deviation, as a measure of the uncertainty of the measured dark counts, is $\sigma_{N_{DCR}} = \sqrt{n_{DCR} t_{meas}}$. This defines the noise floor of the detector if no light is present. The quantity $\sigma_{N_{DCR}}$ is used as the lower limit for calculating the dynamic range [58]. The upper limit for calculating the dynamic range is defined as the maximum possible count rate n_{max}. The dynamic range DR is the ratio of upper and lower limit,

$$\mathrm{DR} = 20\log\left(\frac{n_{max}}{\sqrt{n_{DCR}}} \cdot \frac{t_{meas}}{\sqrt{t_{meas}}}\right) = 20\log\left(\frac{n_{max}}{\sqrt{n_{DCR}}} \cdot \sqrt{t_{meas}}\right). \tag{3.19}$$

Because the relative impact of the detector noise is reduced by averaging, the definition of the dynamic range DR scales with $\sqrt{t_{meas}}$. Low detector noise is key in low-light applications such as fluorescence lifetime measurements. In laser rangefinders, however, low detector noise is not a primary design goal because the detector is operated in the presence of background light and the photon noise dominates (see Section 3.2.2).

3.2.3.5 Afterpulsing

Afterpulses are secondary pulses generated after a primary count. The afterpulse probability (P_{AP}) is the ratio of the rate of secondary pulses and the rate of primary pulses. During avalanche breakdown, the junction is flushed with carriers, some of which may populate traps from lattice defects. These trapped carriers are released depending on their lifetime with a statistically fluctuating delay time [59] and can trigger a secondary pulse correlated with the previous one [44]. Lifetimes of electrons in traps are in the order of tens of ns to μs [59].

Afterpulsing scales with the avalanche current and the duration of the avalanche since the probability of populating a trap increases with the number of carriers present. The external circuitry impacts the avalanche, firstly by limiting the magnitude and the duration of the avalanche current and secondly by the time the SPAD is kept below the breakdown voltage (hold-off time) before recharge. Carriers released during the hold-off time propagate through the (still reverse biased) junction without triggering a self-sustaining avalanche. A long hold-off time will therefore reduce afterpulsing but also increase the dead time. However, a long dead time reduces the maximum possible count rate n_{max} (Eq. (3.18)) and therefore the dynamic range. Nonetheless, in a low-light-level application a high maximum count rate may be sacrificed for a low afterpulsing probability.

The influence of afterpulses on a distance measurement with a laser rangefinder, in particular the influence on the measured time of flight (timing influence), will be analyzed further in Section 3.5.3.

3.2.3.6 Jitter

The jitter is the statistical time variation between true photon arrival and actual SPAD output pulse. The jitter is typically described by the full width at half maximum (FWHM) of a histogram of arrival time differences [35, 60].

There are two main contributions to jitter. The first one is caused by the timing variation of a carrier drifting to the avalanche multiplication region of the SPAD. The drift time depends on where the e-h pair is generated with respect to the multiplication region. The drift time can be estimated with 10 ps/µm [35]. Moreover, photons can be absorbed deeper in the semiconductor and slowly diffuse into the multiplication region. This effect manifests as a slow 'diffusion tail'. As photons with longer wavelength penetrate deeper into the material, the diffusion tail is wavelength-dependent and more prominent in the red and NIR.

The second contribution is the statistical build-up time of the avalanche current. Other than in APDs the avalanche build-up does not result in amplitude noise because the SPAD output is a digital pulse of constant height and width that is determined by the external circuitry. To illustrate this uncertainty, in [61] the timing difference between an avalanche triggered at the center of the active SPAD area and an avalanche triggered at the edge of the active SPAD area of a 40 µm diameter SPAD is given with 200 ps.

There is a tradeoff between jitter and photon detection efficiency. A thick depletion region leads to a higher photon detection efficiency η_{PDE} since the absorption width w_A is increased. However, the jitter is worse since the arrival time of carriers that pass the corresponding larger absorption width w_A, spreads over a larger time interval [62]. The timing uncertainty increases.

The timing influence of jitter in a laser rangefinder will be described further in Section 3.5.1.

3.2.3.7 Supply voltages

Whilst supply voltages for SPAD operation are not key parameters for a laboratory setup, they have significant impact on the cost of electronics in commercial products. Low voltage levels are desired for easy generation within the application-specific integrated circuit (ASIC) for the laser rangefinder. Higher voltages may require additional external components or even an external voltage supply that increases system cost.

Reach-through structures that operate at a breakdown voltage of $V_{bd} = 420\,\text{V}$ and an excess bias voltage of $V_{eb} = 20\,\text{V}$ have been reported by Dautet, McIntyre et al. in [63]. Planar CMOS integrated structures with a rather low breakdown voltage of $V_{bd} = 14.4\,\text{V}$ and an excess bias voltage of $V_{eb} = 1.4\,\text{V}$ have successfully been fabricated by Richardson et al. [33].

3.2.3.8 Array Fill Factor

Multiple SPADs can be implemented as an array as illustrated in Fig. 3–9. Each cell of the array is a SPAD pixel. The pixel pitch defines the grid of the array, i.e., the pixel pitch gives the size of the pixel. The active area of a SPAD is the photosensitive region of a pixel. Only the light falling onto the active area is detected. Each pixel comprises the SPAD device and the surrounding support circuitry for quenching, recharge and routing the signals into and out of the array.

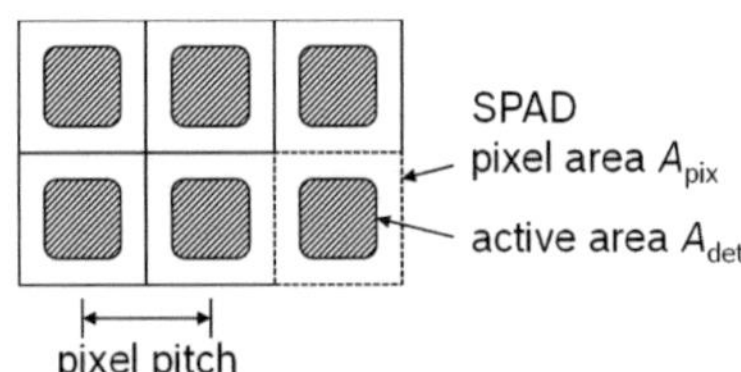

Fig. 3–9: SPAD array

The array fill factor (FF) defines the ratio of the active area of the SPAD for detection and the entire area of a SPAD pixel,

$$FF = \frac{A_{det}}{A_{pix}} \, .$$

(3.20)

There is a tradeoff between large fill factor and jitter. A higher fill factor can be achieved with larger active area SPADs, which come at the price of a higher jitter as explained above in Section 3.2.3.6.

Microlenses are one solution to recuperate fill factor without impairments of the jitter. Microlenses or µlenses are known from image sensors. These lens elements capture light with an aperture larger than the detector active area and direct the light onto the detector active area. Wafer-level microlenses are manufactured directly within the semiconductor process with additional masks and resist layers. With this technology, the diameter of the light sensitive active area of a detector can be increased by about 1 µm. This technology is readily applied in image sensors [64]. As an alternative, microlenses can be fabricated as a part of the chip package. A packaging level microlens can cover the entire cell area of the detector. However, challenges have to be solved in terms of manufacturing and accurately aligning these lens elements on top of SPADs [65].

3.2.3.9 Crosstalk

It is important to distinguish between optical and electrical crosstalk (Fig. 3–10) [35]. Crosstalk is an undesired mutual influence or disturbance.

Electrical crosstalk between SPADs occurs when a SPAD event influences the properties of a neighboring device. For example, an avalanche breakdown in one SPAD can slightly decrease the supply voltage level of neighboring SPADs and thereby reduce the photon detection efficiency of the neighboring SPADs.

The subsequent circuitry is also affected as the voltage levels of decision thresholds shift. Especially when considering fast discharge or recharge times of the SPAD, one has to bear in mind electrical crosstalk effects because of the strong driving currents involved. Moreover, the modulation current of the laser driver in a laser rangefinder can impact the SPAD electrically.

Optical crosstalk: Light from the transmitter can directly impact the receiving SPAD if sending and receiving path are not properly isolated from each other. Moreover, scattering within the optical system or optical package of the chip can introduce optical crosstalk.

SPAD-to-SPAD optical crosstalk occurs, when a SPAD emits photons by electroluminescence during avalanche breakdown. Since silicon is an indirect semiconductor, the probability of a radiative recombination is low. Lacaita reports a photon emission efficiency of 2.9×10^{-5} photons per carrier crossing the pn-junction [66]. However, the self-sustaining avalanche for each SPAD output pulse involves a large number of carriers. In our experiments (see Section 4.2.4) we measured on average $0.53 \times 10^{6} ... 2.0 \times 10^{6}$ electrons per SPAD pulse. The emitted photons can cause an avalanche breakdown of neighboring SPADs which avalanche breakdowns are correlated with the time of the first avalanche. SPADs with large active area have a higher capacitance, and therefore a larger number of carriers is involved in the avalanche. They have a higher probability of a radiative recombination with the emission of a photon. Optical crosstalk of 20% has been measured in [61].

A special case of combined electro-optical crosstalk is caused by a photon which is absorbed deep within the semiconductor below a first SPAD. The absorbed photon generates an electron-hole pair that diffuses laterally towards the active region of a second neighboring SPAD, where an output pulse is triggered in the active region.

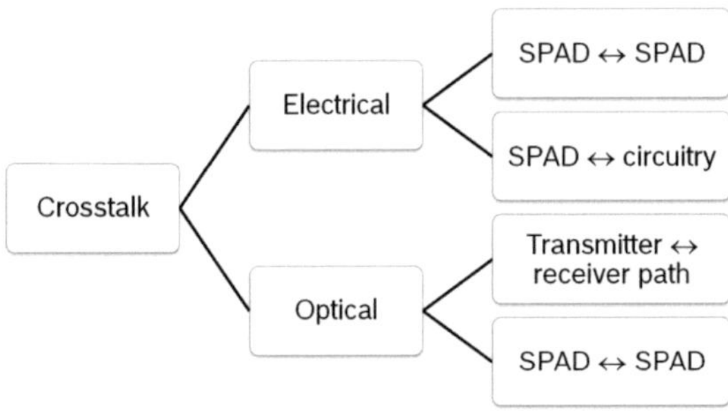

Fig. 3–10: Crosstalk paths in a SPAD-based laser rangefinder

3.3 SPAD Technology

Pioneering work in single-photon detection with solid state photodetectors by McIntyre [67] and by Haitz [68] at the Shockley research laboratory dates back to the 1960s. More details about the history can be found in [9, 35, 69]. The section below focuses on different structures leading to the CMOS SPADs used in this research.

3.3.1 Reach-Through Structures

Reach-through APDs came up as a replacement for photomultiplier tubes because of their insensitivity to magnetic fields and smaller size. The reach-through structures implemented by McIntyre, Petrillo et al. [67, 70, 71] are pn-junctions which are operated at high reverse bias voltage above the breakdown voltage in order to allow single-photon detection.

Fig. 3–11 (a) shows the p+-π-p-n layer sequence from [70], where π denotes a weakly p-doped layer. Most of the light incident on the p+ side is absorbed in the π-layer.

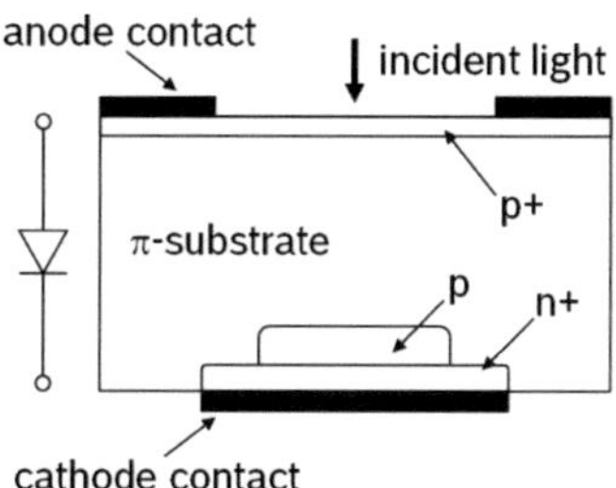

Fig. 3–11: Schematic device structure of reach-through APD [70]. The weakly p-doped substrate is denotes as π-substrate.

The reach-through structure has a long absorption zone that ensures high quantum efficiency. However, the length of the absorption zone defines the voltage which has to be applied to maintain a sufficient electric field in the absorption and multiplication zone (see Section 3.1.2). The p-implant, also referred to as an 'enhancement' implant, lowers the breakdown voltage because it enables reduced field strength in the absorption zone followed by a narrow high-field multiplication region. Nevertheless the 50 % photon detection efficiency (PDE, see Section 3.2.3.1) in [70] comes at the price of a 140 µm thick device with ~400 V breakdown voltage. Reach-through SPADs with up to

65% PDE are commercially available for high-price research instrumentation [72].

While reach-through devices feature good quantum efficiencies, their co-integration capability is very limited. Furthermore the double-sided processing of the semiconductor is a challenging and expensive manufacturing process. In particular the alignment of front- and backside is crucial. Thus a single-sided process (as shown in Fig. 3–12) is to be preferred.

Aull et al. investigate reach-through SPADs for NIR laser ranging systems and 3D cameras [48]. They follow a hybrid-chip approach, with a first chip with reach-through SPADs for photo-detection and a second chip in an RF CMOS process for the timing circuitry. While both chips are optimized for a specific task, the inter-chip connections and the manufacturing process are challenging.

The advantages and disadvantages of SPAD based on reach-trough structures are listed in Table 3.1.

Advantages	*Disadvantages*
⊕ High quantum efficiency ⊕ Timing circuit in RF-optimized process	⊖ High breakdown voltage ⊖ Thick substrate must be free of traps ⊖ Dedicated process with limited co-integration capability ⊖ Challenging manufacturing process ⊖ Cost

Table 3.1: Advantages and disadvantages of reach-trough SPADs

3.3.2 Planar SPADs

The predecessor of planar SPADs is a semiconductor structure designed for the study of breakdown effects in a pn-junction by Haitz [54]. The structure was not yet optimized for light detection but shows the basic structural features found in state-of-the-art planar SPADs.

One of the challenges in planar SPADs is that both the p- and the n-contact are on the same side of the substrate. The active photosensitive area extends in the same plane in lateral direction. A shield structure, the so-called 'guard ring', prevents edge breakdown at the side of the active area and leads to a homogeneous electric field in vertical direction. Fig. 3–12 (a) shows the simplified structure from [54]. The p-contact (anode) is separated from the n+ implant (cathode) by a low-doped n- implant, which reduces the electric field strength below breakdown for the direct lateral anode-cathode path. The high-field zone now homogeneously develops in vertical direction between the p-substrate and the n+ implant. Cova et al. implemented this structure for single-photon detection [73].

Planar SPADs have been further optimized with a strong focus on jitter. In a jitter-optimized structure [74] these SPADs eventually reached 28 ps FWHM (full width at half maximum) jitter at room temperature. Especially the diffusion tail from carriers slowly diffusing into the active region (see Section 3.2.3.6) is reduced by introducing a second pn-junction between p-epistrate and n-substrate [60, 75]. This second pn-junction acts as a protective structure that catches carriers that have been generated deep below the multiplication region and prevents them from triggering a SPAD output pulse. Fig. 3–12 (b) depicts the simplified structure of a dual-junction SPAD, where a p-epistrate is grown onto an n-substrate. Reverse biasing the p-epi/n-substrate-junction draws away carriers that have been absorbed deep within the semiconductor. As an alternative to the guard ring, the structure features an enhancement implant to confine the high-field multiplication region and to prevent edge-breakdown.

While an extraordinarily low jitter is achieved, these structures rely on a dedicated fabrication process with limited yield. Dark count rates are as high as several 100 kHz. The photon detection efficiency does not exceed 16% [62].

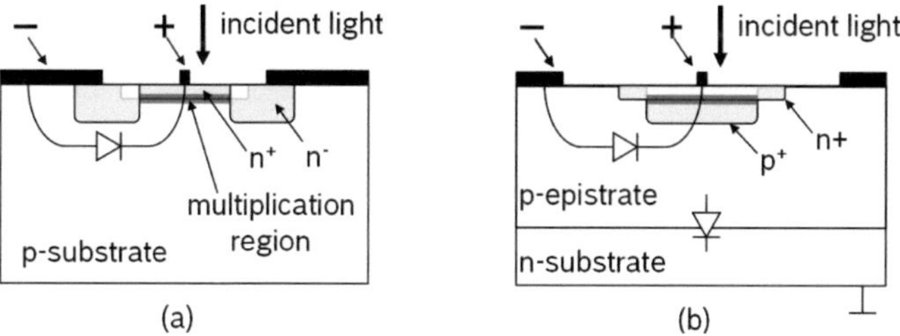

Fig. 3–12: Planar SPAD. (a) The n⁻-implant forms a guard ring to prevent edge-breakdown. A multiplication region with high electric field develops between p-substrate and n⁺-implant. (b) Dual-junction SPAD with additional pn-junction between p-epistrate and n-substrate to reduce the diffusion tail. Instead of a guard ring, a p⁺ implant locally enhances the electric field for a confined multiplication region. Diode schematics illustrate the pn-junctions.

The advantages and disadvantages of planer SPAD that are fabricated in a dedicated fabrication process are listed in Table 3.2.

Advantages	***Disadvantages***
⊕ Low jitter	⊖ Limited quantum efficiency
⊕ Lower breakdown voltage (tens of Volts)	⊖ Dedicated process
	⊖ Cost

Table 3.2: Advantages and disadvantages of
planar SPADs in dedicated fabrication process

The first SPAD fabricated in an industrial CMOS process was published in 2003 by Rochas et al. at EPFL [76]. This is a breakthrough in manufacturing since CMOS SPADs can easily be co-integrated with other circuitry and digital signal processing. Previous SPAD implementations relied largely on off-chip quenching and recharge circuits that can now be easily co-integrated in an industrial grade fabrication process. Moreover, an industrial fabrication process guarantees lower production cost and increased yield.

On the downside, standardized processes provide a limited toolbox of layers, implants and doping profiles. Hence, the SPAD cannot be fully optimized for a specific design goal, such as low jitter, as in case of a dedicated process [60].

The high doping concentrations of implants that are available in nanometer-scale CMOS processes allow the formation of a multiplication region with sufficient field strength for self-sustaining avalanches. Rochas successfully designed a p+-n-well SPAD junction with a p-well guard ring. He only used the elements that are available in the existing 0.8 µm high voltage CMOS process [76].

A first fully integrated SPAD array with readout circuitry was realized in 0.35 µm CMOS [77]. The process was not designed for photon detection efficiency. While the junction design was fine, a passivation layer on top of the SPAD active area limited the photon detection efficiency (PDE) at around only 5%. A follow up design with a slightly modified process reached up to 40 % PDE at 450 nm wavelength and 4 V excess bias voltage [78].

Further designs have been implemented at the 0.18 µm [79], 130 nm [80] and 90 nm [81] process nodes recently. High doping concentrations in nanometer scale CMOS and the respective strong electric fields over short distances allow low breakdown voltages of down to 10 V as demonstrated in [82]. At the same time devices with extremely high doping concentrations and strong electric fields suffer from strong band-to-band tunneling. This gives rise to a high dark count rate (DCR) of hundreds of kHz.

A solution to this DCR problem is presented by Richardson in [33]. Richardson employed so-called well-structures to slightly reduce the peak electric field. The structures are readily available in a 130 nm imaging-process used in mass production of cell-phone cameras. The fabrication process is well-established by STMicroelectronics on an industrial scale. This type of SPAD has been selected for this thesis. The device structure will be shown in more detail in the next section (Section 3.3.3, see Fig. 3–13). On the downside for the timing circuitry of a laser rangefinder, the imaging process has a limited RF-performance.

As an outlook, recent advances in a 90 nm imaging technology by the same research group in cooperation with STMicroelectronics are about to be published. The devices have a median dark count rate (DCR) of only 1 Hz for a small 2 µm diameter disk-shaped SPAD [83] and reach a PDE of up to 20 % in the near infrared [84].

The advantages and disadvantages of SPADs in a standard CMOS process are listed in Table 3.3.

Advantages	*Disadvantages*
⊕ Low breakdown voltage ⊕ Co-integration capability ⊕ Mass-production CMOS process ⊕ Cost	⊖ Medium quantum efficiency ⊖ Limited RF performance of process

Table 3.3: Advantages and disadvantages of planar SPADs in standard CMOS

3.3.3 CMOS SPAD used for Laser Rangefinder

The devices characterized in course of this thesis are CMOS SPADs that are fabricated in a standard imaging process [33, 64] without process modification of the existing process. All structural features are readily available in STMicroelectronics' IMG175 process. The imaging process has several optimizations for high photon detection efficiencies. However, all the structural features of the SPAD device are also available without imaging specific process options. Therefore, these SPADs are also compatible with standard nanometer scale CMOS [33].

Fig. 3–13 shows a schematic cross section through the device. The metal layer of the plus contact connects to a buried n-well implant via n-well and n+ contacts. The buried n-well is formed by high energy ion implantation. It is typically used for isolating NMOS transistors from the substrate. The buried n-well features a retrograde doping profile, i.e., the doping concentration increases with increasing depth. The metal layer of the minus contact connects to a p-well implant via a p+ contact. The multiplication region forms between p-well and buried n-well. The pn-junction of the SPAD is indicated by a SPAD diode D_S and the SPAD capacitance is C_S. The pn-junction that forms between buried n-well and the p-EPI substrate is indicated by a parasitic diode D_P and the parasitic capacitance C_P.

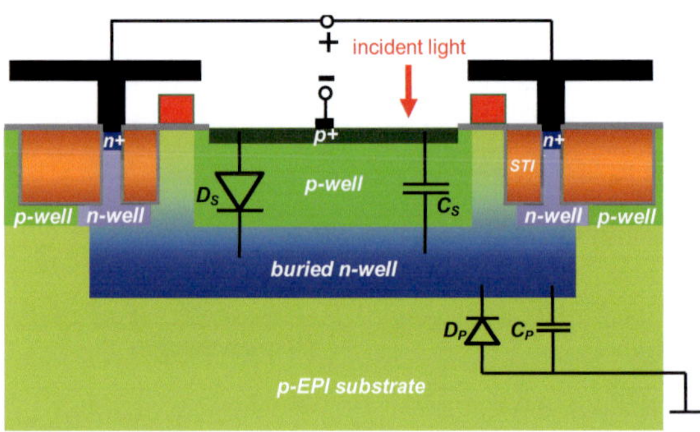

Fig. 3–13: Cross section of a p+/p-well/buried n-well SPAD. SPAD diode D_S, SPAD capacitance C_S, parasitic diode D_P, parasitic capacitance C_P. Modified from [64] © IEEE 2011

In contrast to a direct p+/n-well structure, the lower doping concentrations of the p-well/buried n-well active region slightly weaken the electric field. This structure reduces band-to-band tunneling and significantly lowers the dark count rate (DCR). The broadened active region comes at the price of moderately increased excess bias voltage and a higher jitter. Favorably, the increased junction width improves the wavelength response in the red part of the spectrum. A computer simulation of the electric field can be found in [64].

A SiO_2 shallow trench isolation (STI) provides good isolation in lateral direction. This enables narrow spacing of further circuitry in an array. Thus a low pixel pitch and high fill factor can be realized. Since STI induces strain in the lattice and increases the number of charge traps, which in turn increases

DCR and afterpulsing [85], the STI isolation is separated from the central p-well of the SPAD.

An imaging technology has some peculiar properties and optimizations [37] that can be seen as either additional benefits or drawbacks compared to a regular CMOS process. The optical stack, i.e., the layers above the photodetector, is optimized for transmission of light. Instead of several metal layers for routing, the number of metal layers is limited to only one or two for anode and cathode supply voltages. In-pixel circuitry for active quenching and routing of control signals may require extra metal layers. Further optional in-pixel circuitry, for example for selectively disabling SPADs to reduce power consumption in a handheld device, requires additional wiring effort.

The laser rangefinder ASICs implemented for this work feature three metal layers and therefore do not fully leverage the full photon-detection efficiency (PDE) potential of the imaging technology. There is a tradeoff between pixel pitch and photon detection efficiency. A high PDE can come at the price of a larger pixel pitch. With a reduced number of metal layers the optical stack is shorter, and less light is absorbed. However, supply connections have to be routed in parallel in the same layer which increases the area consumption. Alternatively, connections can be routed in multiple stacked metal layers at the price of a higher optical stack and thus a lower PDE.

Moreover, not only the number of layers but also layer thickness and layer spacing are reduced in this technology. On the one hand, this improves optical transmission; on the other hand, these layers with reduced thickness have a lower conductivity. Furthermore, the thin layer spacing has detrimental effects on the RF performance. The closely spaced metal layers act as a 'parallel plate capacitor' with high parasitic capacitance. Further information about the optical stack and process steps, which are known to reduce the trap density, can be found in [35].

Besides the cost for development effort and mask manufacturing, silicon area is the driving cost-factor in ASICs for a low-cost laser rangefinder. The small features size in this 130 nm process favorably reduces the area occupation of the digital and signal processing part, and also reduces the required area for the front-end circuits.

3.4 Front-End Circuits

This chapter describes the circuits directly adjacent to the SPAD for quenching and recharge as well as for generating the SPAD output pulse. These so-called 'front-end circuits' are mandatory for SPAD operation firstly because they prevent damaging the device with high avalanche currents due to the self-sustaining avalanche and secondly to reset the SPAD for detection of another photon. Front-end circuits can be grouped into passive quenching (PQ) circuits, Section 3.4.1, and active quenching (AQ) circuits, Section 3.4.2. The front-end circuits also determine the saturation behavior of the SPAD, Section 3.4.3.

3.4.1 Passive Quenching

A passive quenching circuit is as simple as a series resistance. Fig. 3–14 (a) shows a SPAD in series with a resistor R_Q. V_{bd} is the breakdown voltage of the SPAD, V_{eb} the excess bias voltage at which the SPAD is operated above breakdown. V_S is the floating voltage of the SPAD node.

The transient behavior of photon detection, avalanche quenching and recharge is depicted in Fig. 3–14 (b). At time t_0 a photon strikes the SPAD and generates an electron-hole pair which initiates an avalanche. The avalanche current i_S builds up and causes a voltage drop across the quenching resistor R_Q. The voltage V_S decreases. When V_S drops below breakdown threshold voltage V_{br} of the device the avalanche is not self-sustaining anymore and eventually stops. Then, the SPAD has a high impedance again and can be recharged to V_{eb} through the same series resistance R_Q that has been used for quenching. The quantity $t_{90\%}$ denotes when the SPAD floating node voltage has recharged to 90% of V_{eb}.

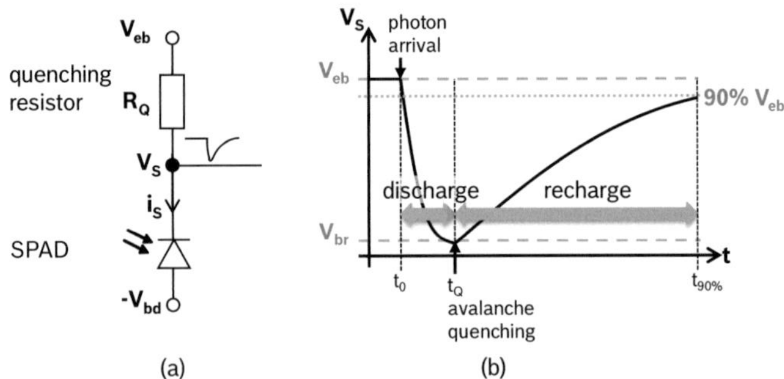

Fig. 3–14: Passive quenching circuit. Excess bias voltage V_{eb}, breakdown voltage V_{bd}, breakdown threshold V_{br}, quenching resistance R_Q, current through SPAD i_S, SPAD floating node voltage V_S. (a) schematic (b) voltage curve at SPAD floating node.

This behavior can be described with the model by Haitz [68], where the SPAD is simplified as shown in Fig. 3–15. C_p denotes the parasitic capacitance and C_S denotes the SPAD capacitance that have to be discharged and recharged. The device switches between low and high (infinite) resistance. Upon photon arrival the switch S closes. The capacitors are discharged through the low-resistance path R_S. R_Q is large compared to the SPAD series resistance R_S during breakdown. After quenching the switch S opens, so that the junction capacitance and the parasitic capacitance are slowly recharged via the quenching resistor.

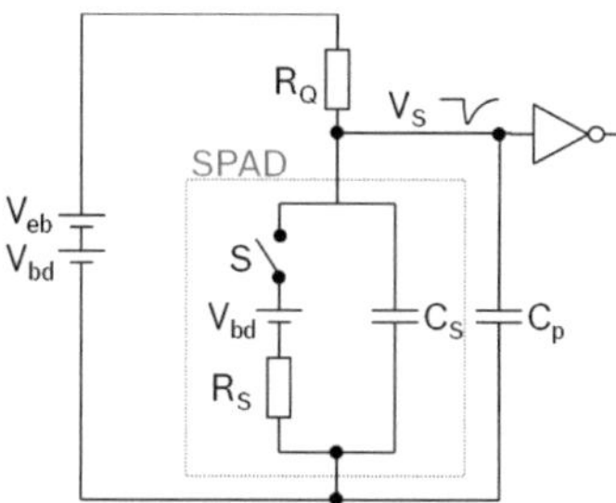

Fig. 3–15: Simplified SPAD model with passive quenching circuit [35, 68]. Excess bias voltage V_{eb}, breakdown voltage V_{bd}, SPAD floating node voltage V_S, quenching resistance R_Q, SPAD series resistance R_S, switch S.

Recharging the SPAD is governed by the RC time constant from the quenching resistor R_Q and the parallel diode capacitance C_S and parasitic capacitances C_p. For CMOS SPADs, the quenching resistor is realized as a MOS transistor, shown in Fig. 3–16 (a). The resistance and thereby the recharge time can be controlled via the gate voltage V_{quench} to recharge times from typically several tens to hundreds of nanoseconds.

An inverter connected to the SPAD floating node in Fig. 3–16 (a) senses the voltage drop at V_S. With only two transistors, a CMOS inverter is a very area efficient device. The CMOS inverter determines the decision threshold voltage V_{th}. The idealized output of the inverter is a digital pulse of constant height and width. In reality the inverter has a limited slew rate and supply noise and electrical crosstalk affect V_{th}. Also pixel-to-pixel mismatch of V_{th} may have to be considered for accurate timing at picoseconds time scale.

Fig. 3–16 (b) shows the voltage at the SPAD floating node V_S over time. The CMOS inverter switches to high logic level, when the SPAD crosses the threshold voltage V_{th} at time 0. The dead time (see Section 3.2.3.3) in this context is the time until V_S surpasses V_{th} again at time t_{dead}. A higher recharge resistance R_Q will increase the RC time constant for recharge and thereby increase the dead time t_{dead}.

A short dead time allows for a higher maximum count rate that is ultimately limited by $n_{\text{max}} = 1/t_{\text{dead}}$ (Eq. (3.18)). However, there is a tradeoff between afterpulsing (see 3.2.3.5) and count rate. The appropriate dead time depends on the application or even within an application from the operating conditions. For a laser rangefinder operated at low background light, a low count rate and therefore long dead time will be sufficient because of the limited received photon rate. This way the impact of afterpulsing is reduced. At high background light, however, noise from background light can be dominant over impairments due to afterpulsing. In that case a high count rate and short dead time is desirable.

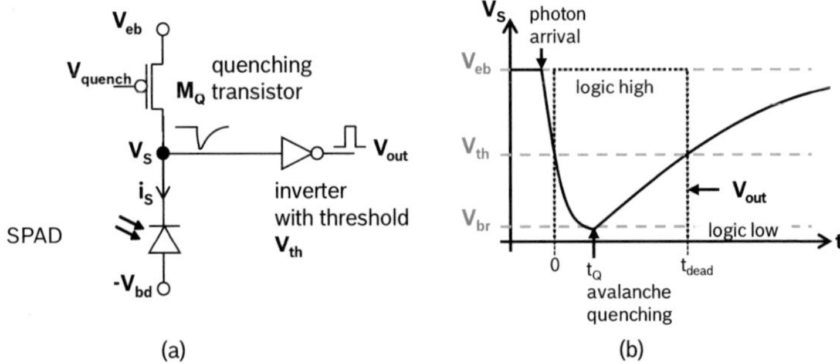

Fig. 3–16: Passive quenching circuit. (a) Schematic with quenching transistor and comparator. (b) Voltage curve at SPAD floating node V_S. V_{th} denotes the decision threshold of the inverter. V_{out} is the voltage of the output of the comparator.

It should be noted that the SPAD is already biased above breakdown voltage shortly after quenching, but yet before surpassing the decision threshold V_{th}. Another avalanche breakdown can occur. The effect of these sub-threshold events on count rate will be described in Section 3.4.3 in the context of detector saturation and paralysis.

The photon detection efficiency (PDE, see Section 3.2.3.1) scales with the excess bias voltage. Therefore the PDE recovers with increasing V_S until finally reaching V_{eb}. Even though after t_{dead} the SPAD is biased above V_{th}, the SPAD does not see quite as many photons yet as if it were fully biased to V_{eb}. Hence, right after the dead time a slightly decreased PDE is expected.

3.4.2 Active Quenching

An active quenching circuit operates in three phases: (1) active discharge, (2) hold-off and (3) recharge. Fig. 3–16 (a) shows a schematic of an active quenching circuit. After detecting a photon, the control (not shown) of the active quenching circuit starts a positive feedback loop and closes switch M_Q at $t = 0$. This current path quickly discharges the SPAD and pulls the SPAD floating node V_S to ground. The SPAD is held at ground for a period of time called hold-off time. The hold-off time is required to reduce afterpulsing (see 3.2.3.5). Thereafter the switch M_Q is opened (not conducting) and switch M_R is closed (conducting). The SPAD quickly recharges through this low-impedance path. Upon reaching V_{eb} or after a predefined recharge time, the switch M_R is opened again. Fig. 3–16 (b) sketches the transient behavior.

Active quenching presents several benefits over passive quenching in all three phases:

(1) After sensing the initial avalanche onset, M_Q provides a second, alternative current path for discharge which reduces the charge flow through the SPAD. Firstly, a reduced afterpulse probability is expected because the probability of populating traps reduces. Secondly, a reduced charge flow reduces optical crosstalk by photoluminescence. Thirdly, a reduced charge flow could also reduce power consumption (however this is not expected in practice because of the power consumption of the additional circuitry for active quenching and recharge).

(2) Trapped carriers that are released or photons incident during the hold-off time cannot start an avalanche event because the SPAD is biased below breakdown during the entire hold-off time.

(3) Because of the strong recharge, the SPAD quickly reaches its full PDE. This increases the count rate.

On the downside, active quenching requires additional circuit elements with the respective area occupation, not only for the circuit elements themselves but also for wiring. Additional metal layers may be required. This increases the height of the optical stack which in turn reduces the PDE. Moreover, the high peak-currents for fast discharge and recharge can cause electrical crosstalk.

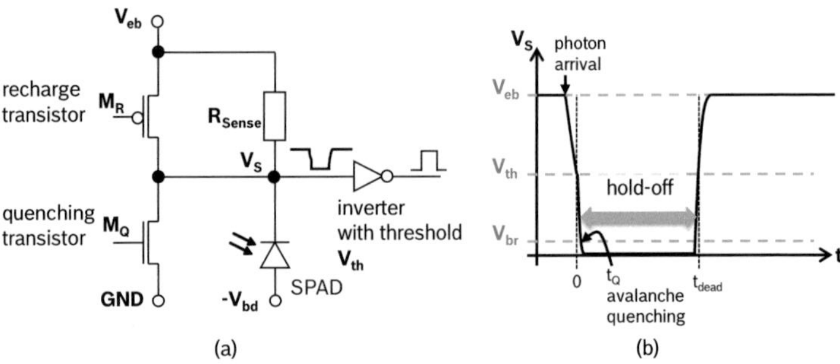

Fig. 3–17: Active quenching circuit. (a) Schematic with actively controlled quenching and recharge transistor. (b) Transient voltage curve at SPAD floating node V_S. V_{th} denotes the decision threshold of the inverter.

One implementation of an active quenching circuit will be shown and explained below with reference to Fig. 3–18. This circuit has been proposed theoretically in [86]. It was implemented and published as part of this research with a record values for count rate and dynamic range [53].

The switches M_1 and M_3 are normally on, i.e., closed (conducting / low resistance) for a logic low level at their gates and open (not conducting / high resistance) for a logic high level. The switches M_2 and M_4 are normally off, i.e., open for a logic low at their gates and closed for a logic high.

Upon photon arrival the avalanche current causes a voltage drop across M_3 and reduces the voltage at the SPAD floating node V_S. The positive feedback loop closes switch M_1 and pulls midpoint to V_{eb}. The logic high closes switch M_4, which pulls V_S to ground. The SPAD floating node is quickly discharged to ground. At the same time the logic high at midpoint propagates through a delay line of inverters. An additional control signal (not shown) can tune rise and fall times of the inverters and thereby control the delay time that sets the hold-off time. After the hold-off time, a logic low reaches and closes switch M_3. Just one gate delay later a logic low closes switch M_2, which in turn pulls midpoint to logic low and stops active quenching from transistor M_4. The SPAD is actively recharged through the low-impedance path M_3. The logic low at midpoint propagates through the delay chain and eventually opens switch M_3. Both M_3 and M_4 are then opened (not conducting / high resistance). The SPAD is rearmed and ready to detect another photon. Dead times down to 5.6 ns have been realized with this circuit.

Further information about quenching circuits is provided by Cova et al. in [44].

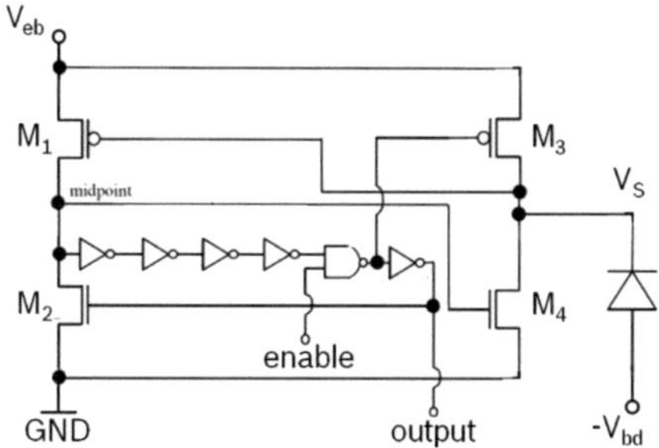

Fig. 3–18: Active quenching circuit implemented as part of this research [53]. The hold-off time of this monostable circuit is defined by the inverter chain.

3.4.3 Saturation and Paralysis

An ideal photon counter should not be influenced by previous detection events. Any photon should be detected regardless of photon rate or time since the last event. At sufficiently long times between subsequent events, the detector has fully recovered. But what happens at short inter-arrival times, i.e., high count rates? The nature of this recovery is determined by the type of quenching circuit.

Fig. 3–19 shows the transient voltage curve of the SPAD floating node with additional photons incident during the dead time (a) for a passive quenching circuit and (b) for an active quenching circuit. (a) In a passive quenching circuit, an avalanche can be triggered by another photon as soon as V_{br} is exceeded. As a consequence, V_S can remain below the threshold voltage V_{th} of the inverter for an extended period of time. This extends the dead time and "paralyzes" the detector. As long as $V_S < V_{th}$, no further SPAD output pulses are generated that can be seen by the subsequent circuitry. The SPAD is 'blind'.

(b) In an active quenching circuit the SPAD is held at ground during the entire hold-off time. A second photon incident during hold-off time therefore cannot trigger a second avalanche event and is simply disregarded. As a consequence the dead time is not extended, leaving the actively quenched SPAD "non-paralyzable" during hold-off. However, at very high photon rates the count rate of the actively quenched SPAD eventually saturates when the SPAD is triggered again and again right after the dead time is over. As previously mentioned, the count rate of a SPAD is ultimately limited by the dead time by $n_{max} = 1/t_{dead}$ (Eq. (3.18)).

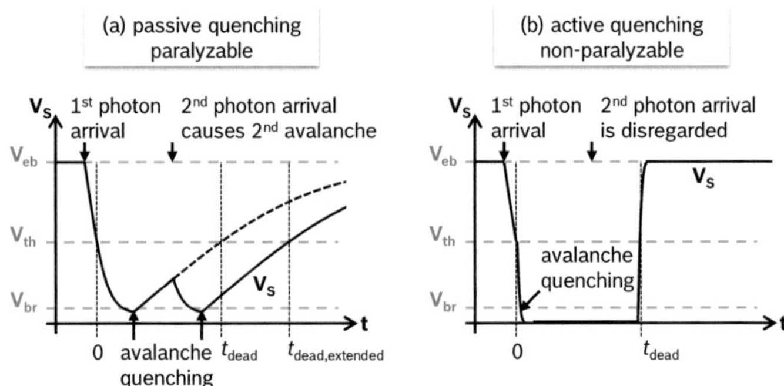

Fig. 3–19: SPAD voltage V_S and dead time t_{dead}. (a) Another photon incident during dead time can extend the dead time of the passive quenching circuit to $t_{dead,extended}$. (b) The active quenching circuit does not suffer from dead time extensions.

While the 2nd photon is not counted in both quenching circuits, the dead time or the extended dead time have an impact on the count rate. In the following, the counting behavior of a SPAD with quenching circuit is described analogous to a Geiger-Müller counter with models adopted from nuclear-particle physics instruments [87].

3.4.3.1 Saturable Model

For active quenching circuits, the relation between actual count rate n (see Eq. (3.15)), including saturation effects, and the ideal count rate m, without saturation effects, is predicted by a saturable-detector model. The ideal count rate m increases linearly with the incident photon rate.

The relation can be derived as follows. The ideal number of counts M that could be measured in an measurement time t_{meas} equals the number of the actually measured counts N plus the missed counts M_{missed},

$$M = N + M_{missed} \ .$$ (3.21)

The missed counts M_{missed} are the part of the ideal number of counts M that fall into the dead times t_{dead} that follow the actually measured counts N. For each measured count N the active quenching circuit starts a hold-off time equal to the dead time. Thus the total dead time during the measurement time t_{meas} is $N t_{dead}$. In consequence, $N t_{dead}/t_{meas}$ is the fraction of the measurement time and thus the fraction of the ideal number of counts M that does not contribute to the actual counts N. Equation (3.21) can therefore be rewritten as

$$M = N + \frac{N t_{\text{dead}}}{t_{\text{meas}}} M \ .$$
(3.22)

The ideal count rate m can be written as

$$m = \frac{M}{t_{\text{meas}}} = \frac{N}{t_{\text{meas}}} + \frac{N}{t_{\text{meas}}} t_{\text{dead}} \frac{M}{t_{\text{meas}}}$$
$$= n + n t_{\text{dead}} m \ .$$
(3.23)

Solving Eq. (3.23) for the actual measured count rate $n(m)$ as a function of the ideal count rate m gives

$$n(m) = \frac{m}{1 + m t_{\text{dead}}} \ .$$
(3.24)

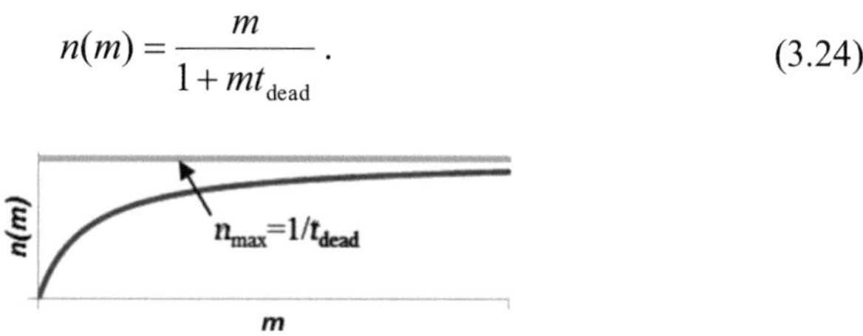

Fig. 3–20: SPAD count rate with saturable detector model.
Ideal count rate m, true count rate n.

For high ideal count rates $m \to \infty$, the actual count rate eventually saturates at $n_{\text{max}} = 1/t_{\text{dead}}$.

3.4.3.2 Paralyzable Model

For passive quenching circuits, the paralyzable detector model gives a first order approximation between actual count rate n and the ideal count rate m for a paralyzable detector such as a passively quenched SPAD. The derivation relies on the inter-arrival time, i.e., time difference in the arrival time of subsequent Poisson distributed photons (see Section 3.2.2). The derivation is based on [87].

An event only causes a valid count, if no other event has occurred during the dead time t_{dead} prior to detection. Only those counts which arrive at time intervals τ_2 greater than the dead time $\tau_2 > t_{\text{dead}}$ are recorded. As given in Section 3.2.2, the probability density of inter-arrival times τ_2 of events with an event rate m is

$$p(\tau_2) = m\exp(-m\tau_2). \tag{3.25}$$

The probability P that $\tau_2 > t_{\text{dead}}$ is

$$P(\tau_2 > t_{\text{dead}}) = \int_{t_{\text{dead}}}^{\infty} m\exp(-m\tau_2)d\tau_2 = \exp(-mt_{\text{dead}}). \tag{3.26}$$

The actual number of counts N that is observed during the measurement time t_{meas} is just the fraction of the ideal number of counts M whose inter-arrival times satisfy the condition

$$N = M\exp(-mt_{\text{dead}}). \tag{3.27}$$

The actual count rate $n(m)$ as a function of the ideal count rate m for a paralyzable detector is then given by

$$n(m) = m \cdot \exp(-mt_{\text{dead}}). \tag{3.28}$$

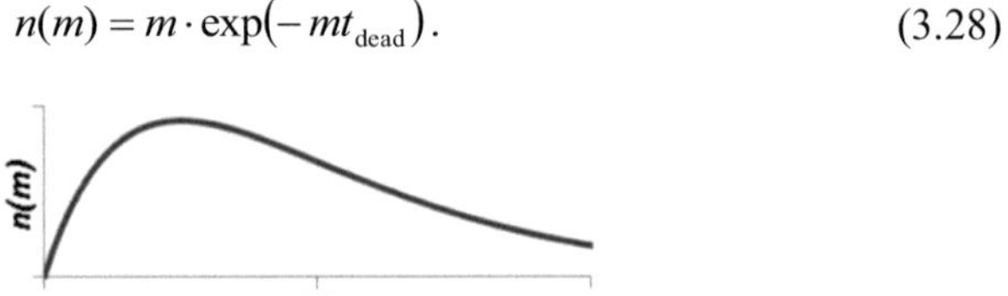

Fig. 3–21: SPAD count rate with paralyzable detector model.
Ideal count rate m, true count rate n.

For high ideal count rates $m \to \infty$, the negative exponent dominates and eventually paralyses the SPAD completely, $n \to 0$. The maximum count rate $n_{\text{max}} = 1/(et_{\text{dead}})$ is reached at $m = 1/t_{\text{dead}}$.

3.4.3.3 Hybrid Model

The hybrid detector model is a combination of the saturable and the paralyzable detector model [88, 89]. Fig. 3–22 shows the behavior of a detector that follows the hybrid detector model. The dead time t_{dead}, as the minimum time the SPAD takes to be ready to detect another photon (see Section 3.2.3.3), comprises a saturable (non-paralyzable) dead time portion and a paralyzable dead time portion, $t_{\text{dead}} = t_{\text{dead,saturable}} + t_{\text{dead,paralyzable}}$. A photon incident during the saturable dead time, denoted by an arrow with a diamond in Fig. 3–22, does not extend the dead time and is simply disregarded. Only photons incident during the paralyzable dead time, denoted by an arrow with a filled circle in Fig. 3–22, extend the dead time.

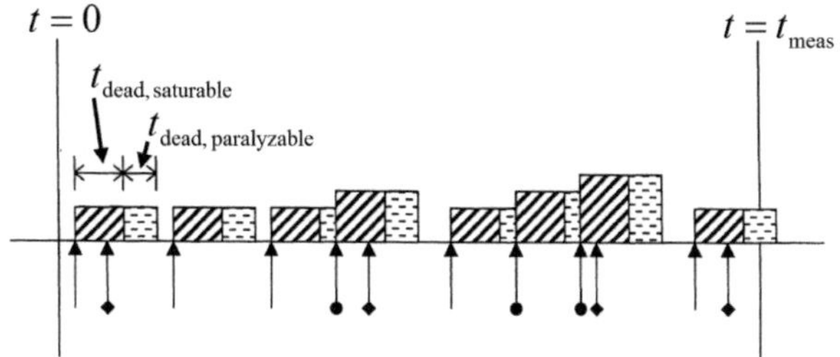

Fig. 3–22: Hybrid detector. Saturable dead time $t_{\text{dead,saturable}}$, paralyzable dead time $t_{\text{dead,paralyzable}}$, measurement time t_{dead}. Arrows denote incident photons. Arrows with a diamond denote photons incident during the saturable dead time which do not extend the dead time. Arrows with a disk denote photons incident during the paralyzable dead time which do extend the dead time. © Elsevier 2000 [88]

This model is appropriate for SPADs because both active and passive quenching circuits can feature a paralyzable and a non-paralyzable section for the following reasons:

In passive quenching circuits, a second photon incident during avalanche breakdown does not cause paralysis, because the junction is already being flushed with carriers. This introduces a saturable (non-paralyzable) element $t_{\text{dead,saturable}}$ to passive quenching. During the passive recharge phase, the SPAD is paralyzable as described above. This introduces a paralyzable element $t_{\text{dead,paralyzable}}$.

For active quenching circuits, the circuit is non-paralyzable during avalanche breakdown and during the entire hold-off time since the SPAD is biased below the breakdown voltage. This defines the saturable (non-paralyzable) contribution $t_{\text{dead,saturable}}$ to the dead time. In active quenching circuits the recharge phase is critical. A strong recharge transistor can push the floating node voltage V_S over the decision threshold V_{th} even when the SPAD fires during the recharging process. Such a device is entirely non-paralyzable. However, an active quenching circuit with a weak recharge transistor can be paralyzed during the short period of time $t_{\text{dead,paralyzable}}$ that is required for recharging the device from V_{br} above threshold V_{th}.

Equation (3.29) gives the count rate for a device having time intervals where it is saturable and time intervals where it is paralyzable. The derivation can be found in [88].

$$n(m) = \frac{m \cdot \exp\left(- m t_{\text{dead,paralyzable}}\right)}{1 + m t_{\text{dead,saturable}}} \tag{3.29}$$

For $t_{\text{dead,paralyzable}} = 0$, Eq. (3.34) corresponds to the saturable model in Eq. (3.24). For $t_{\text{dead,saturable}} = 0$, the Eq. (3.34) corresponds to the paralyzable model in Eq. (3.28).

The maximum count rate n_{max} depends on the contributions of the saturable and the paralyzable dead time. It can be calculated using Eq. (3.34) at

$$m_{\text{max}} = \frac{- t_{\text{dead,paralyzable}} + \sqrt{t_{\text{dead,paralyzable}}^2 + 4 t_{\text{dead,paralyzable}} t_{\text{dead,saturable}}}}{2 t_{\text{dead,paralyzable}} t_{\text{dead,saturable}}}. \tag{3.30}$$

Niclass and Soga presented a circuit with a remarkably low dead time of 6 ns [58]. Despite deploying active recharge, the circuit is partially paralyzable and reaches a maximum count rate of 66 MHz, i.e., 40% of the theoretical maximum count rate of $n_{\text{max}} = 1/6\,\text{ns} = 167\,\text{MHz}$ for a purely saturable quenching circuit.

Fig. 3–23 shows the actual count rate n as a function of the incident photon rate $p = m / \eta_{\text{PDE}}$ at $\eta_{\text{PDE}} = 20\%$ and $t_{\text{dead}} = 20\,\text{ns}$ for the count rate models presented above. As a limitation it should be noted that neither of the models considers the reduced photon detection efficiency when the floating node voltage V_S is still recharging towards V_{eb}.

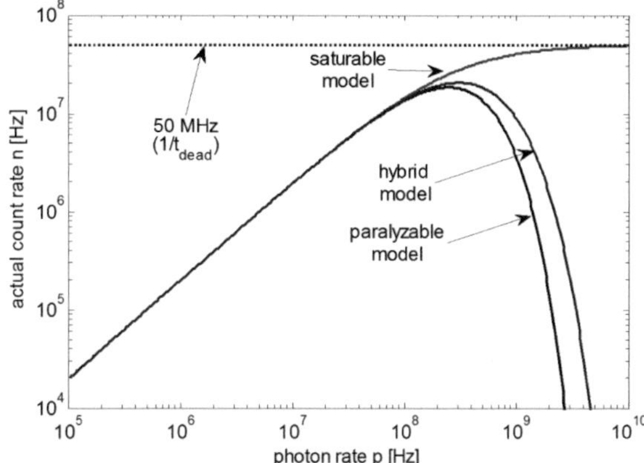

Fig. 3–23: SPAD count rate n including saturation effects as a function of the photon rate p. Photon detection efficiency $\eta_{\mathrm{PDE}} = 20\%$ and dead time $t_{\mathrm{dead}} = 20\,\mathrm{ns}$. As a first approximation, actively quenched SPADs follow a saturable detector model and passively quenched SPADs follow a paralyzable detector model. The hybrid detector model accounts for both saturable and paralyzable periods during a quenching/recharge cycle (here: 10 ns saturable dead time and 10 ns paralyzable dead time).

3.4.3.4 Effect on Laser Rangefinder

A laser rangefinder for measuring distances based on frequency-domain reflectometry (see Section 2.2.3) receives an approximately constant background light intensity I_{BL} and modulated signal light I_S with a CW intensity modulation on top on the background light (see Eq. (2.7)). This corresponds to a photon rate p_{BL} due to background light and a modulated photon rate p_S due to signal light. The numbers of photons acquired during a measurement time t_{meas} are N_{BL} and N_S respectively. It should be noted that measured photon numbers include saturation effects of the detector.

For the background light dominated case, i.e., $p_{BL} \gg p_S$, and assuming Poisson distributed photons, the photon noise can be approximated by the standard deviation of the number of measured photons due to background light $\sigma_N = \sqrt{N_{BL}}$ (see Section 3.2.2). For low relative photon noise, as defined in Eq. (3.13), the number of measured photons and thus the photon rate should be as high as possible.

The SNR of a laser rangefinder for the background light dominated case can be approximated by the ratio of counts from signal light N_S and the photon noise,

$$\text{SNR} = \frac{N_S}{\sqrt{N_{BL}}}.$$
(3.31)

However, also saturation effects have to be taken into account. For the background light dominated case, wherein the signal light is a small modulation on top of the background light, the transfer function of the ideal signal light counts to true signal light counts is essentially the slope of the respective curve in Fig. 3–23. In terms of the ideal count rate m and actual true count rate n, the impact of saturation effects on the signal modulation can be described by the differential modulation efficiency η_{mod},

$$\eta_{\text{mod}}(m) = \frac{dn(m)}{dm}.$$
(3.32)

At low count rates, the modulation efficiency is $\eta_{\text{mod}} = 100\%$ since saturation effects can be neglected at low count rates and the actual count rate corresponds to the ideal count rate. The modulation efficiency decreases towards higher count rates as saturation effects set in.

Including saturation effects, the SNR of a laser rangefinder results from a trade-off between modulation efficiency (η_{mod} goes down at high count rates) and count rate (relative noise goes down at high count rates). In terms of the ideal count rate m, this relation is given by

$$\text{SNR}(m) \propto \frac{\eta_{\text{mod}}(m)}{\sqrt{n(m)}}.$$
(3.33)

From this relation an optimum operating point for a laser rangefinder can be derived. The relation from Eq. (3.33) is plotted in Fig. 3–24. The curves for paralyzable and hybrid model are zero at the inflection point of the count rate graphs in Fig. 3–23. Increasing the photon rate above the maximum count rate will actually decrease the measured count rate. With the given values of $\eta_{\text{PDE}} = 20\%$ and $t_{\text{dead}} = 20\,\text{ns}$, this system has its optimum operating point at an actual count rate of around $14\,\text{MHz}$ for paralyzable and hybrid model and $24.5\,\text{MHz}$ for the saturable model. The maximum achievable SNR is more than

10% higher for the saturable model than for the hybrid or paralyzable model at the same dead time.

As a solution to reach both a low relative impact of photon noise and high modulation efficiency, one large SPAD can be split into an array of several small devices [21]. Each small SPAD operates at a lower count rate far from saturation. The counts of the plurality of SPADs are added in the digital domain. The price to pay is a lower fill-factor since each SPAD requires additional control circuitry.

To save at least some chip area for routing several lines out of an array, several SPAD output signals of the SPADs of the SPAD array can be combined on a bus. However, if the output signals of multiple SPADs are combined on a bus, the information which pulse belongs to which SPAD is lost. For using a SPAD array in a time-of-flight distance measurement system, it is therefore a premise for combining multiple SPADs onto a bus that the average time between the arrival of a photon and the time of providing an output pulse has to be the same for all SPADs of the array. This requires very good matching of the individual SPADs (and support circuitry) in fabrication.

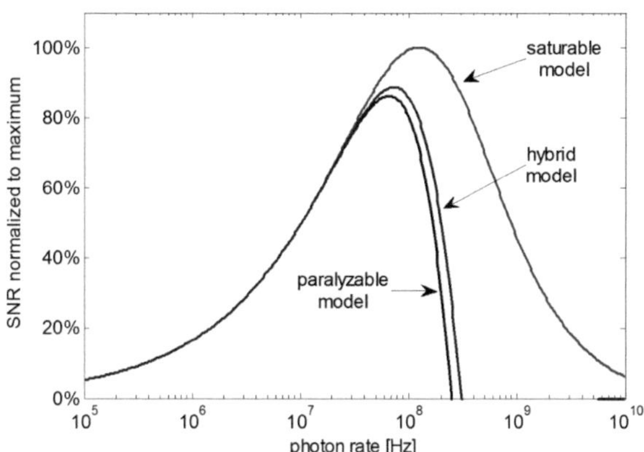

Fig. 3–24: SNR as a function of the photon rate. Photon detection efficiency $\eta_{PDE} = 20\%$ and dead time $t_{dead} = 20\,\text{ns}$. Non-paralyzable, actively quenched SPADs following the saturable detector model feature a more than 10 % higher maximum SNR and offer a broader operating range at high SNR than SPADs following the paralyzable or hybrid model.

Saturation not only has an influence on count rate but also on the temporal characteristic of a SPAD. This effect has not been paid much attention to in the prevalent low light applications of SPADs [34]. The influence on the temporal characteristic of a SPAD will be discussed in Section 3.5.4.

3.4.4 SPAD Drive Circuit Options

There are two drive circuit options for the SPAD in a front-end circuit: (a) negative-drive option (ND) or (b) positive-drive option (PD).

In negative-drive option Fig. 3–25 (a), the SPAD is biased with the anode connected to negative drive voltage $-V_{bd}$. The SPAD floating node potential varies between V_{eb} and ground with "negative" pulses at V_S. The SPAD capacitance C_S and the parasitic p-substrate / n-well capacitance C_p are discharged and recharged with every event. The parasitic p-substrate / n-well diode D_p only has to withstand voltages up to V_{eb}. Please refer to Fig. 3–13 for the device structure which also shows C_S, C_p and D_p.

In positive-drive option Fig. 3–25 (b), a high positive voltage $V_{HV} = V_{eb} + V_{bd}$ is applied to the SPAD cathode. The SPAD floating node varies between ground and V_{eb} with "positive" pulses at V_S. The advantage of positive-drive option over negative-drive option is that only the SPAD capacitance C_S is discharged and recharged with every event. The parasitic capacitance C_p remains at a constant potential. This implies a reduced charge flow through the SPAD with the respective benefits on power consumption, afterpulsing and fast timing. Using the PD option the parasitic p-substrate / n-well diode D_p has to endure a voltage of V_{HV}. Otherwise D_p breaks down before the SPAD. This effect is called premature parasitic breakdown.

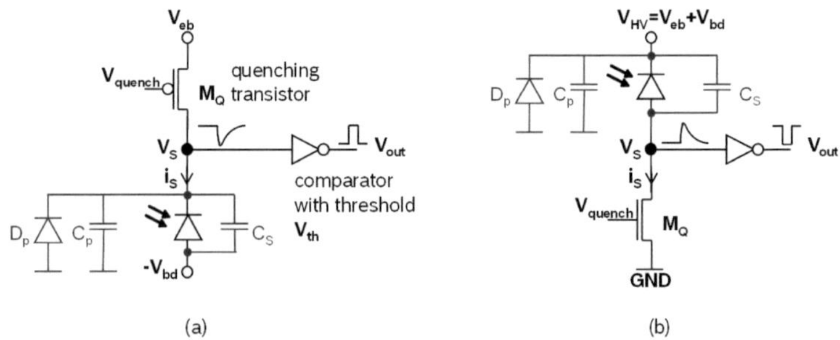

Fig. 3–25: SPAD drive circuit options. (a) negative-drive (b) positive-drive. In negative-drive option the parasitic capacitance C_P is discharged and recharged with each SPAD event but not in positive-drive option. In positive-drive option the parasitic diode D_p has to withstand a higher voltage.

3.5 Temporal Characteristic of a SPAD

The target of this research is a laser rangefinder that measures a distance based on time of flight of a signal from the rangefinder to the target and back. Thus the temporal characteristic of a SPAD requires special attention and is highlighted in this section.

An ideal detector for a time-of-flight laser rangefinder should detect the correct arrival time of a photon independent of any previous detection events. Furthermore, the time between photon arrival and output pulse should be perfectly constant regardless of the temperature and the intensity of the incident light. Unfortunately this is not true for a real-world device. The reasons are explained below.

3.5.1 Jitter

Jitter has previously been defined in Section 3.2.3.6 as the statistical time variation between true photon arrival time and actual appearance of the SPAD output pulse. Jitter is typically in the range of several tens of picoseconds FWHM (full with at half maximum) for special timing-optimized dedicated structures [74]. For CMOS integrated SPADs [64], jitter is typically in the range of hundred to a few hundreds of picoseconds FWHM.

A laser rangefinder must resolve the time of flight with a timing resolution of 6.7 ps for a distance resolution of 1 mm (see Section 2.2). A single-shot distance measurement, where only one SPAD pulse is evaluated, suffers from the entire SPAD jitter and can therefore not achieve millimeter accuracy. However, for multiple single-shot distance measurements, the time of flight can then be determined from a histogram of arrival times, e.g., by evaluating the position of the peak or by calculating the mean arrival time. Therefore, even though the SPAD jitter is significantly broader than the required timing resolution, the jitter does not necessarily limit the performance of the laser rangefinder.

The jitter of a SPAD can be seen similar to the dithering of an audio signal before analog-to-digital (A/D) conversion. Dithering adds amplitude noise with zero mean onto an analog signal. Without dither, the analog signal would be assigned one digital level. With dither, the received signal is distributed over more than one digital level. From the frequency of occurrence at the different digital levels, the average value of the received signal can be determined. Depending on the integration time, the resolution of this value is finer than the quantization levels of the A/D converter. Analogously, the SPAD jitter can distribute the received signal over multiple sampling windows (bins) over time. The average arrival time of the SPAD pulses can be calculated from the histogram.

3.5.2 Pile-Up Effect

With "pile-up" we denote a distortion of the jitter for large photon count rates [90]. The time variation between a photon generating an electron-hole pair and the actual SPAD output pulse largely depends on the build-up of the avalanche current. At very high intensities there is a significant probability of generating multiple e-h pairs within the time scale of SPAD jitter that would be measured at low photon rates.

Fig. 3–26 shows an exponential build-up of the avalanche current i_A as a function of the time t. A first photon causes a first avalanche current (avalanche 1); a second photon causes a second avalanche current (avalanche 2). The combined avalanche current, denoted by avalanche 1+2, is twice as large and crosses the detection threshold by a time difference $\Delta t_{\text{pile-up}}$ earlier than the current from an individual avalanche. In a laser rangefinder, however, a shorter time of flight corresponds to a shorter distance. Pile-up can thus cause a distance measurement error.

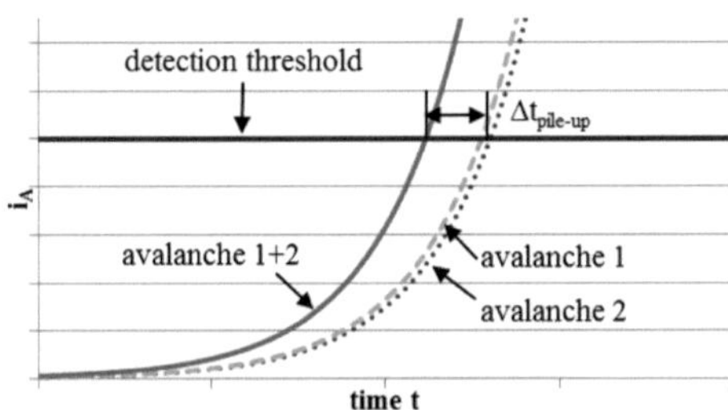

Fig. 3–26: Build-up of the avalanche current i_A as a function of time. A first photon causes a first avalanche current (avalanche 1); a second photon causes a second avalanche current (avalanche 2). The combined avalanche current, denoted by avalanche 1+2, crosses the detection threshold by a time difference $\Delta t_{\text{pile-up}}$ earlier than the current from an individual avalanche.

To estimate the relevance of the pile-up effect for a laser rangefinder, a jitter of 200 ps (FWHM) is assumed. The count rate for having on average one event during the jitter FWHM of 200 ps is $1/200\,\text{ps} = 5\,\text{GHz}$. This count rate is way beyond the expected count rates in a hand-held laser rangefinder (see Section

5.3.1) with a small receiver lens with some millimeters aperture and with a band-pass filter that blocks most of the background light (even at full 100 klux background light from direct sunlight). However, the pile-up effect has to be considered when designing an experiment for characterizing the SPAD jitter. For an undisturbed measurement the test signal, e.g., from a short-pulse laser, must be attenuated such that typically less than one photon per laser pulse is incident on the device under test.

3.5.3 Afterpulsing

Afterpulses are defined in Section 3.2.3.5 as secondary pulses generated after a primary pulse. Instead of one primary pulse whose arrival time corresponds to the true time of flight for the measured target distance, there can be secondary pulses at later times. A pulse at a later time, however, corresponds to a longer target distance.

In systems where the time of flight is measured in the frequency domain (see Section 2.2.3), the influence of afterpulsing on the measured distance depends on the modulation period, the afterpulse probability and the afterpulse distribution.

The effect is described analytically for a simplified case: Each primary pulse at time $t_0 = 0$ is followed by an afterpulse at time t_{AP} later with a probability Prob_{AP}. The impulse response $h_{AP}(t)$ of the detector (including afterpulses, but without SPAD jitter) is

$$h_{AP}(t) = \delta(t_0) + \text{Prob}_{AP}\delta(t_0 - t_{AP}) . \tag{3.34}$$

The received signal intensity I_S without background light is given by

$$I_S(t) = 1 + \cos\left(2\pi f_m t\right) . \tag{3.35}$$

Effectively, the received signal is convolved with the impulse response $h_{AP}(t)$ of the detector,

$$\begin{aligned} I_D(t) &= I_S(t) * h_{AP}(t) \\ &= \left(1 + \cos\left(2\pi f_m t\right)\right) * \left(\delta(t_0) + \text{Prob}_{AP}\delta(t_0 - t_{AP})\right) \end{aligned} \tag{3.36}$$

In frequency domain, the convolution corresponds to a multiplication,

$$F(I_D(t)) = F(I_S(t)) F(h_{AP}(t))$$

$$= \left(\delta(0) + \frac{1}{2}(\delta(f + f_m) + \delta(f - f_m)) \right)\left(1 + \mathrm{Prob}_{AP} e^{-j2\pi f t_{AP}}\right). \qquad (3.37)$$

The time shift of the afterpulse in time-domain translates into a phase shift in frequency-domain [91]. This phase-shift in turn can lead to a distance error as explained below.

The simplified case of afterpulses that occur a fixed time after the primary pulse is illustrated in Fig. 3–27. The SPAD receives a sinusoidally intensity modulated signal with a modulation frequency of 1 GHz, Fig. 3–27 (a). For each photon, the SPAD is assumed to generate one primary pulse at time $t_0 = 0$ with a probability of 100% and an afterpulse at time $t_{AP} = 20.25\,\mathrm{ns}$ later with a probability $\mathrm{Prob}_{AP} = 25\%$, Fig. 3–27 (b). Higher order afterpulses, i.e., afterpulses of afterpulses, are neglected. The resulting detected signal (red curve), shown in Fig. 3–27 (c), consists of a superposition of the primary sinusoidal wave (blue curve) at 1 GHz frequency plus a secondary sinusoidal wave (green curve) at same frequency with 25 % amplitude and a phase shift of $\pi/2$. The system can only detect the phase of the combined signal, which includes a phase error caused by the secondary wave.

The maximum phase error for an afterpulses probability of $\mathrm{Prob}_{AP} = 25\%$ is 245 mrad. At a modulation frequency of 1 GHz, this phase error corresponds to a distance error of 5.8 mm.

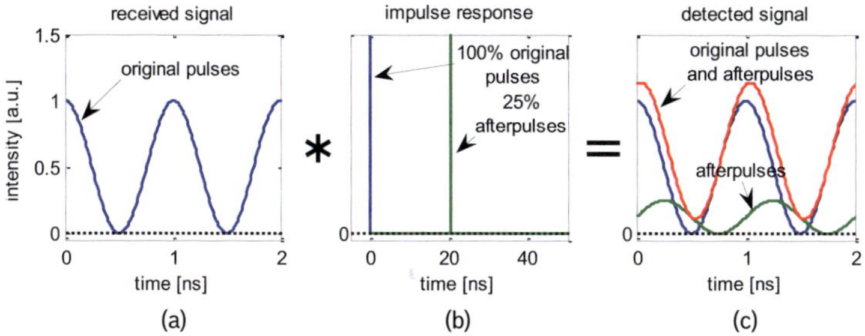

Fig. 3–27: Influence of afterpulses on the detected signal. Segment of a continuous sinusoidally intensity modulated signal with a frequency of 1 GHz. (a) Original pulses (blue) that would be caused by a detector without afterpulsing. (b) Impulse response of the detector. Afterpulses occur 20.25 ns after the original pulse with an afterpulse probability of 25%. (c) Detected signal. Afterpulses cause a second sinusoidal wave with 25% of the amplitude of the primary signal. The secondary wave is delayed by 20.25 ns which corresponds to a phase shift of $\pi/2$. The detected signal (red) with afterpulses is a superposition of the signal from original pulses (blue) and afterpulses (green).

Fortunately, afterpulses are not single pulses at a fixed time after the primary pulse. Trapped charges are released over time. The distribution of afterpulses is governed by an exponential decay [59]. Decay constants are in the range of tens of ns to several µs.

The case of exponentially distributed afterpulses is illustrated in Fig. 3–28. In this example, the afterpulse probability is 25% and the first afterpulse is also detected 20.25 ns after the primary pulse, as shown in Fig. 3–28 (b). However, the afterpulses are exponentially distributed with a decay constant of 20 ns. The resulting detected signal (red curve), shown in Fig. 3–28 (c), consists of a superposition of the primary sinusoidal wave (blue curve) at 1 GHz frequency plus a contribution from afterpulses (green curve). However, the afterpulse contribution gives slightly more than an offset. This can be explained by the impulse response of the afterpulses that is wider than several modulation periods which acts as a low-pass filter. In contrast to the exponentially distributed afterpulses in Fig. 3–28 (b), the Dirac impulse response of the afterpulses in Fig. 3–27 (b) does not act as a filter.

For a modulation frequency of 1 GHz, exponentially distributed afterpulses with a decay constant of 20 ns and an afterpulse probability of 25 %, we obtain a maximum phase error of 2 mrad. This corresponds to a distance error of 48 µm.

At a low modulation frequency of 30 MHz, however, the distance error caused by the same impulse response is 1.5 mm.

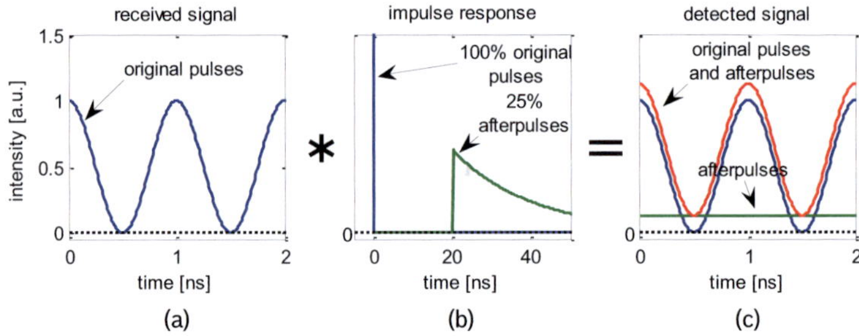

Fig. 3–28: Influence of afterpulses on the detected signal. Segment of a continuous sinusoidally intensity modulated signal with a frequency of 1 GHz. (a) Original pulses (blue) that would be caused by a detector without afterpulsing. (b) Impulse response of the detector. Afterpulse probability of 25%. Exponential decay of the afterpulse probability with a decay constant of 20 ns. (c) Detected signal. Because the decay constant is long compared to the modulation period of 1 ns, the impact on the phase of the detected signal (red) is very limited. The impact of the afterpulses reduces to negligibly more than an offset (green).

In conclusion, a laser rangefinder can tolerate a high degree of afterpulsing if the afterpulsing is distributed over several RF periods. The impact of afterpulsing, especially for time-of-flight distance measurement in frequency domain, depends not only on the absolute afterpulse probability but also on the afterpulse distribution and on the signal modulation frequency. Favorably, the modulation frequency is selected much larger than the decay constant.

The fact that even a high afterpulsing probability can be tolerated is a peculiarity of the distance measurement application. For fluorescence lifetime measurements, a prominent application area for SPADs, a very low afterpulsing is a key design parameter.

3.5.4 Quenching

The quenching process influences the temporal characteristic of the SPAD both during discharge as well as during the recharge phase.

A prerequisite for identical timing is that all events start from the same voltage level V_{eb}. This ensures the same delay (apart from jitter) between the time a photon strikes the detector to the time where the resulting voltage crosses the decision threshold V_{th}. This effect has to be considered in high light level applications, where there is a non-negligible probability of a photon striking the SPAD before the SPAD has fully recharged (see Section 3.4.1). If the avalanche starts from a lower voltage level, the time to crossing the decision threshold V_{th} is shorter. Thus apparently a shorter distance is measured.

The relevance of this effect can be estimated for the passively quenched SPAD in [76] which has an RC time constant of $\tau_{RC} = 25.7\,\text{ns}$ ($R_q = 270\,\text{k}\Omega$, $C = 95\,\text{fF}$). A passive recharge circuit requires $\tau_{50-90\%} = 41.4\,\text{ns}$ time to recharge the SPAD from $V_S = 50\%$ V_{eb} up to 90% V_{eb}. The count rate where there is on average one event incident during this recharge time is $n = 1/41.4ns = 24.2\,\text{MHz}$.

With active recharge, the recharge time can be reduced since the SPAD is quickly recharged to V_{eb} through a low resistance path. The probability of a photon striking the SPAD before it has been biased to V_{eb} is dramatically reduced.

Nonetheless, there is a problem with active recharge as well, if the recharge current provided by the recharge transistor exceeds the avalanche current. As described previously, such a recharge transistor can push the floating node voltage V_S over the decision threshold V_{th} even when the SPAD fires during the recharging process. In this case the avalanche is already burning during recharge. Nevertheless the SPAD floating node voltage is increased towards and then held at V_{eb} until active recharge stops and the recharge transistor opens (not conducting). Only then the floating node voltage breaks down below the decision threshold V_{th}. In this case, the SPAD output pulse represents the instant in time when active recharge stopped, but not the arrival time of the photon triggering the avalanche. This is illustrated in Fig. 3–29.

Fig. 3–29 shows the voltage V_S at the SPAD floating node over time. Blue arrows denote photon arrivals. Green arrows denote the time, when a corresponding output pulse should be detected at a fixed time after the photon arrival. This fixed time between photon arrival and should-be pulse is the time required to discharge the SPAD from the excess bias voltage V_{eb} below the decision threshold V_{th}. Red arrows denote the time of the actual output pulse that can then be processed by further circuitry. For the first photon, the actual output pulse occurs at the same time t_1 when the output pulse should occur. Hence, the time of the output pulse represents the time of photon incidence.

Fig. 3–29 also shows a second photon that is incident during active recharge. As described above, the recharge transistor provides enough recharge current to pull the SPAD floating node voltage V_S to and then hold it at V_{eb} even when the avalanche is burning. Only after active recharge has ended, the avalanche current causes V_S to drop below the decision threshold V_{th}. This is when an output pulse can be detected by the subsequent circuitry. When a photon strikes the SPAD during active recharge, the actual output pulse does not represent the arrival time of the second photon any more. Instead, the time t_2 of the second output pulse represents the end of the active recharge process.

The end of the active recharge process, however, is directly linked to the arrival time of the primary pulse at time t_1. These pulses a fixed time after the primary pulse have the same detrimental effect as afterpulses that occur at a fixed time after the primary pulse as described in the previous section on afterpulsing (see Section 3.5.3, Fig. 3–27).

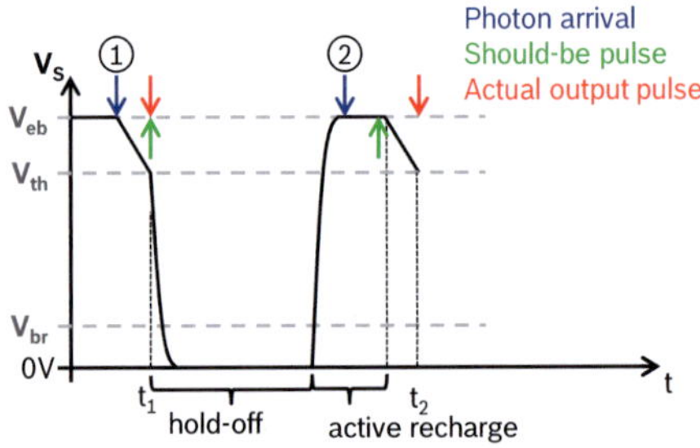

Fig. 3–29: Arrival time of output pulses for SPAD with active recharge. SPAD floating node voltage V_S as a function of time. Blue arrows denote photon arrivals. Green arrows denote the time, where an output pulse should be detected. Red arrows denote the time of the actual output pulse that can be processed by further circuitry. Counts from photons incident during active recharge are released a fixed time after the active recharge has ended.

Technical solutions for this problem include sensing circuits that only allow the detection of a new event a certain delay time after the SPAD has fully been recharged. This again would introduce a paralyzable dead time section (see Section 3.4.3.2) to the SPAD. The only solution for not losing arrival time information during quenching and recharge is probably such a sensing circuit which comes at the price of losing some photons.

As active quenching circuits quickly recharge a SPAD in about 1-5 ns, compared to 41 ns for the passive recharge circuit in the example given above, the probability of photons incident during active recharge is low for the expected event rates in a laser rangefinder. It should be noted that there is a design tradeoff between very short active recharge and electrical crosstalk (see Section 3.2.3.9), because a short recharge time requires high peak currents.

Compensation algorithms have been proposed to recover the true event rate from the measured count rate in [89]. While true event rates can be reconstructed to some extent, these algorithms do not recover timing information.

3.5.5 SPAD Drive Circuit Options

A small device capacitance is key for fast discharge/recharge and accurate timing. Using the positive-drive option, Fig. 3–25 (b), only the diode capacitance has to be discharged. Therefore, steep-edged slopes and an improved timing performance are expected for SPADs that are operated in positive-drive option.

3.5.6 Inhomogeneity

Inhomogeneity in this context refers to a difference between multiple SPADs, in particular multiple SPADs in a SPAD array. The temporal inhomogeneity becomes a problem when signals from multiple SPADs are evaluated.

For example in the context of SPAD saturation, a combination of multiple neighboring SPADs as array can be beneficial as shown in Section 3.4.3. Line lengths in an array of SPADs as well as the load at the SPAD output node can be equalized by design. A careful matching of the parasitics is also required. In practice, this requires extensive simulations in the design process of the ASIC.

Further examples for spatially separated SPADs are the SPADs of target and the reference path in a laser rangefinder. In order to avoid mechanic elements such as a reference flap (see Section 2.3.1), both paths should be measured in parallel. A strictly symmetric layout of the ASIC ensures equal operating conditions for the target and the reference path.

Within the same die a similar behavior can be expected for all devices, i.e., a given property will exhibit temperature drift in the same direction for all transistors of one type. Nonetheless, there will be a residual difference. The purpose of the measures above is to reduce absolute timing errors to differential timing errors that are only affected by the mismatch of otherwise almost identical paths. Despite all of these efforts, however, there will still remain a timing difference due to process variation and doping gradients across the die.

92

Conclusions about SPADs and their suitability for a laser rangefinder will be drawn following the experimental characterization at the end of Chapter 4.

4 SPAD Characterization

This chapter focuses on measurements of top-level SPAD parameters relevant for laser rangefinders. The target is to evaluate the feasibility of a ranging system with these structures.

In a first step existing SPADs from STMicroelectronics are analyzed, always keeping in mind the question of how each parameter eventually influences the performance of a laser rangefinder. In a second step test structures with beneficial properties for a SPAD-based laser rangefinder are implemented in silicon on a further multi-project wafer.

4.1 Experimental Results MPW2009

4.1.1 Test Structures

There are two test chips implemented on the multi-project wafer 2009 (MPW2009). Chip 1 (SPADDEVEL) consists of various single SPADs and 3×3 arrays with access to individual SPAD outputs. These structures have been designed and fabricated by STMicroelectronics. More information about these devices can be found in the PhD thesis of Richardson [35].

Chip 2 (CF340 MPW2009, ASIC 1) includes a 3×3 SPAD array from chip 1 with some additional control logic for an automated measurement setup. The CF340 MPW2009, designed by Bosch Automotive Electronics, also contains means for time-of-flight distance measurement that will be described in Chapters 5 and 6. The SPADs are connected to this means for distance measurement so that there is no simultaneous access to the individual SPAD outputs. Microlenses (Section 3.2.3.8) have not been applied in the test chips.

4.1.1.1 SPADDEVEL

Fig. 4–1 shows a photograph of the SPADDEVEL test chip in an optical LQFP48 package. SPADDEVEL consist of various test structures per die. Different bond variants give access to these structures.

Fig. 4–1: Photograph of SPADDEVEL test chip in an optical LQFP48 package. Single SPAD test cells and 3×3 arrays are implemented on the same die. Different bonding schemes access the different test structures.

The first group of test structures evaluates different sizes and shapes. Fig. 4–2 shows the family of SPADs implemented with diameters ranging from 2 µm to 32 µm. SPAD shapes are disk-shaped, square and square with rounded edges. A comparison of different SPAD types for the SPADDEVEL test chip is given in [35]. The characterization in this work is focused on performance parameters of particular relevance to ranging applications and a system-on-chip ranging system. The measurements will be presented for an 8 µm diameter disk-shaped SPAD.

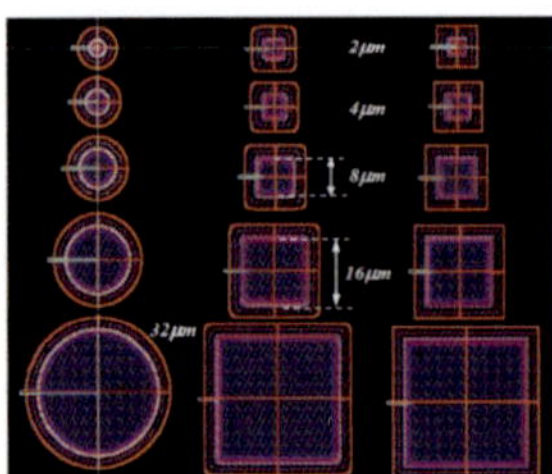

Fig. 4–2: SPAD family. Left column disk-shaped SPADs, center column square SPADs with rounded corners, right column square SPADs. (reprinted from [35])

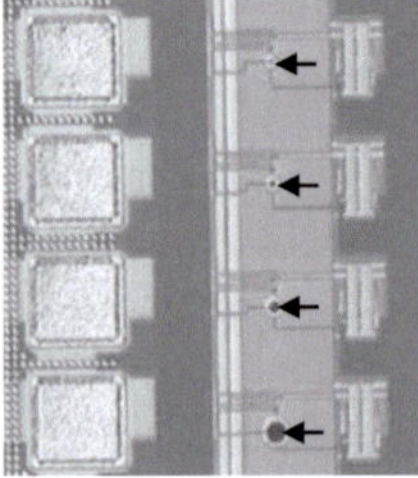

Fig. 4–3: Photomicrograph of four single-SPAD test cells from the size / shape structures. Arrows denote four SPAD active areas of different size.

Each test cell of the size/shape structures (Fig. 4–3, Fig. 4–4) consists of a SPAD with quenching transistor M_{quench} for passive quenching and a CMOS inverter as the threshold comparator. The output of the inverter V_{out} leaves the test chip through an output pad, which is directly contacted by an oscilloscope probe. For direct measurement of the breakdown voltage, the single SPAD test cells also include a test path via M_{test} to bypass the quenching transistor. The quantities V_{eb} and V_{bd} are the applied excess bias and breakdown voltage of the SPAD; V_{quench} the control voltage of the quenching transistor; V_{DD} and V_{SS} the supply voltages of the inverter; $V_{\text{DCTestGND}}$ and V_{DCMODE} the ground connection and control voltage of the bypass.

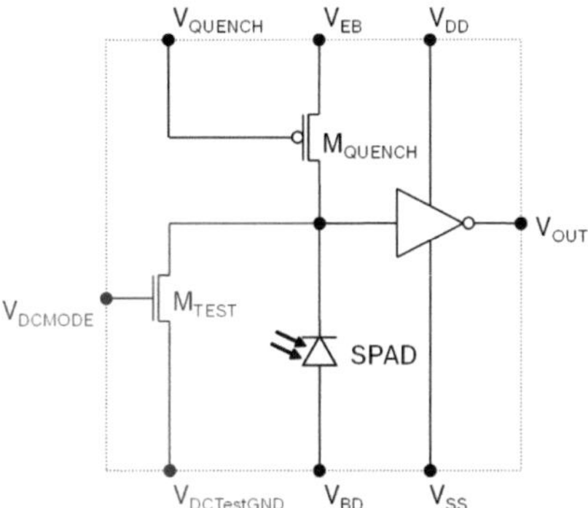

Fig. 4–4: Single SPAD test cell

The second group of test structures comprises 3×3 SPAD arrays. These structures are implemented to analyze the mutual influence of neighboring SPADs. All individual outputs are routed to different output pads. It should be noted that there is only one connection for V_{eb} and one for V_{bd} common to all SPADs. The devices feature passive quenching circuits.

The third type of test structures is a single SPAD with the active quenching circuit presented in Section 3.4.2. A photomicrograph is shown in Fig. 4–5.

A further test structure with SPADs in positive-drive option (see Section 3.4.4) was not functional because of the parasitic p-substrate/n-well-diode (Fig. 3–25, Fig. 3–13) breaking down before actual SPAD breakdown.

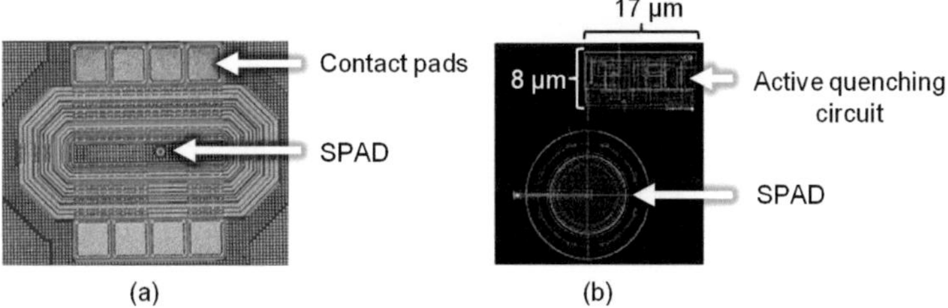

Fig. 4–5: Photomicrographs of (a) active quenching cell, (b) actively quenched disk-shaped SPAD with 8 µm diameter and associated 8 µm × 17 µm active quenching circuit.

4.1.1.2 CF340 MPW2009

The CF340 MPW2009 is the first system-on-chip laser rangefinder that includes a 3×3 SPAD array, laser driver and timing and control circuitry. The CF340 MPW2009 has been designed by Bosch Automotive Electronics for this research. The 3×3 array consists of 8 µm diameter, disk-shaped active area, passively quenched SPADs with negative-drive option from the SPADDEVEL test chip.

Control logic allows automated readout and switching of different SPADs either to an internal counter or a test output pin. Any measurement will include the influence of the respective multiplexers and buffers involved in the signal chain to the output pad. There is no direct access to individual SPAD outputs. This chip is used for automated measurements of the photon detection efficiency (PDE, see Section 3.2.3.1) and the dark count rate (DCR, see Section 3.2.3.2) as well as the temporal characteristic of multiple SPADs.

For comparable results between all test structures, all results will be given for the disk-shaped SPAD with 8 µm diameter unless stated otherwise.

Fig. 4–6 shows a photomicrograph of the 3×3 array implemented in the CF340 MPW2009. SPAD active areas can be clearly seen as the dark light-absorbing disks. It should be noted that there is only one connection for V_{eb} and one for V_{bd} common to all SPADs. SPAD outputs are routed out of the array individually and combined with a programmable multiplexer.

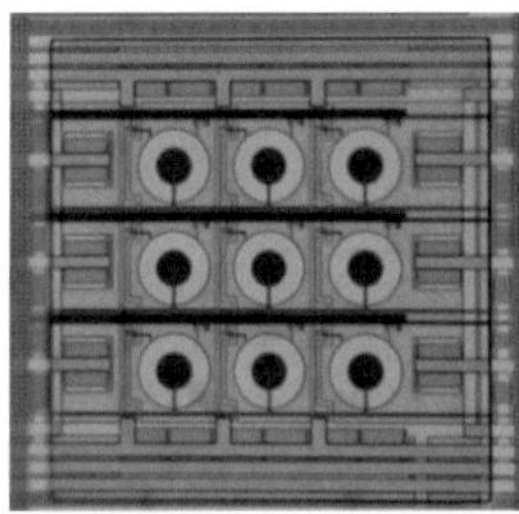

Fig. 4–6: Photomicrograph of CF340 MPW2009 3×3 SPAD array.
The dark disks are the light absorbing active areas with 8 µm diameter.

4.1.2 Performance Parameters

4.1.2.1 Current-Voltage-Curve

Fig. 4–7 depicts the current-voltage-curve (IV-curve) of a reverse biased SPAD pn-junction. The static IV-curve is measured through the bypass path via M_{test}. The breakdown voltage V_{bd} is determined as the threshold where the measured current passes $-i_D \geq 20\,\text{nA}$.

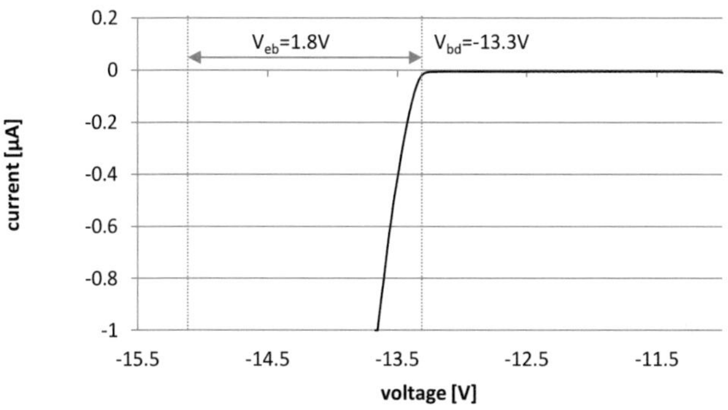

Fig. 4–7: Measured IV-curve

The variation of V_{bd} between different SPADs is of interest for a laser rangefinder in which one large-area APD is replaced by an array of SPADs. The sample-to-sample variation is also of interest since the supply voltage circuitry of a laser rangefinder must cover the entire span of possibly relevant breakdown voltages. Fig. 4–8 depicts V_{bd} of five 3×3 arrays. V_{bd} is consistently around -13.3 V. For each sample, the span from minimum to maximum value is shown. Mean values are connected with a solid line. The span between the maximum and the minimum value for all SPADs of all samples is 0.1 V.

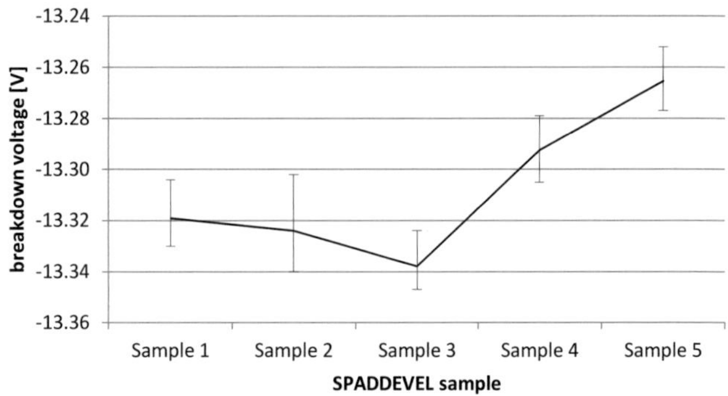

Fig. 4–8: Breakdown voltages for multiple samples of 9 SPADs each.
Mean values and span from minimum to maximum measured value are shown.

As a laser rangefinder is used on a construction site indoors and outdoors, the SPADs must function over a temperature range from -10°C to 50°C. The temperature behavior of V_{bd} is shown in Fig. 4–9. The experimental setup with a Peltier element as the temperature control unit limits the lower experimental measurement range to -5°C. A linear fit estimates the influence of temperature on V_{bd} with 7.9 mV/K.

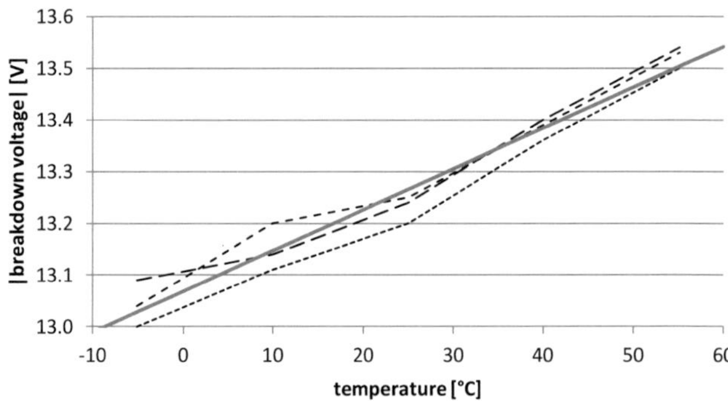

Fig. 4–9: Temperature dependence of V_{bd}.

The increase of $|V_{bd}|$ with temperature is qualitatively in line with theory. At higher temperatures the mean free path length of a carrier reduces. Therefore a higher electric field is required to accelerate the carriers within a shorter path to an energy sufficient for impact ionization. In a final system-on chip rangefinder, a breakdown voltage monitor and a voltage control circuit are to be implemented to track and set the operating point.

4.1.2.2 Photon Detection Efficiency

Photon detection efficiency (PDE, see Section 3.2.3.1) as a function of the wavelength is shown in Fig. 4–10. At higher wavelengths the PDE decreases because less photons are absorbed within the active region of the SPAD. At short wavelengths a significant percentage of the photons is absorbed within the optical stack and does not reach the pn-junction of the SPAD. Carriers generated close to the surface can recombine before reaching the multiplication region [61]. Distortions around 500 nm are due to the optical stack having layers of different refractive index which act as a dielectric filter [35].

Silicon SPADs are well suited for operation in the visible spectrum. SPADs can also be used in the near infrared for 'invisible' ranging applications, however, the PDE decreases towards the infrared. In a laser rangefinder, we explicitly want a visible wavelength since the visible laser beam constitutes both the aiming aid and the measuring medium.

For the measurement evaluation, dark counts are subtracted from the measured counts to isolate counts triggered by photons. PDE is measured at least one order of magnitude below saturation limits.

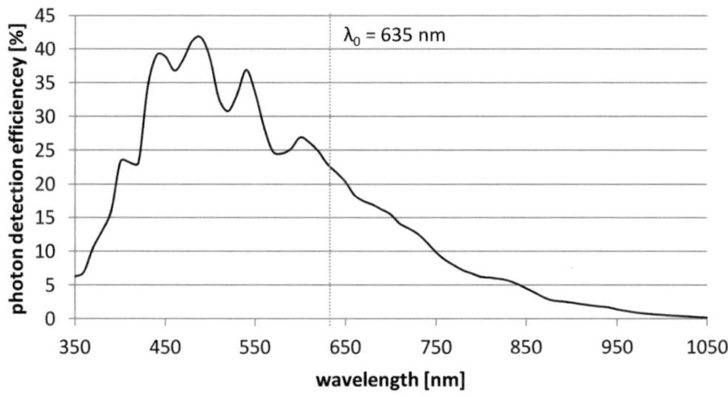

Fig. 4–10: Photon detection efficiency as a function of the wavelength at 1.8 V excess bias voltage. (Measurement data by courtesy of J. Richardson, University of Edinburgh / STMicroelectronics)

This graph shows that a blue or green laser diode is the best choice for high detection efficiency. In the target system, red laser diodes will be used despite the technical disadvantage of a lower PDE because they are available at much lower cost. Therefore, all following measurements will be performed with a red laser diode at 635 nm wavelength, or with an LED centered at 638 nm.

The photon detection efficiency depends on the operating voltage $V_{eb} + |V_{bd}|$ of the SPAD. Fig. 4–11 exemplarily shows the PDE for a voltage sweep of the excess bias voltage V_{eb}. The PDE ranges from 10 % to 26 %. The PDE will eventually saturate when nearly each absorbed photon generates a self-sustaining avalanche.

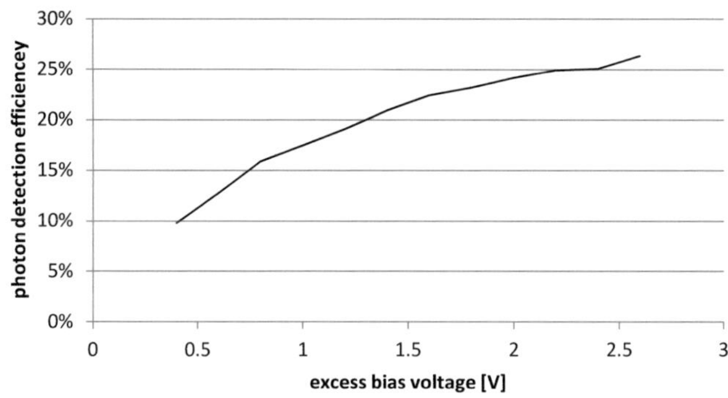

Fig. 4–11: Photon detection efficiency over excess bias voltage.

By controlling the excess bias voltage V_{eb}, the PDE can be easily adjusted. At high illumination levels from strong background light or from a highly reflective target at short distance, V_{eb} can be reduced to prevent saturation or paralysis (see Section 3.4.3). However, the temporal behavior (see Section 3.5) has to be kept in mind. Changing V_{eb} can impact the time from photon absorption to crossing the threshold V_{th}. Therefore, V_{eb} should not be changed during a distance measurement. A quick preliminary measurement to determine an appropriate operating point from the count rate can be followed by a subsequent highly accurate distance measurement at fixed V_{eb}.

Fig. 4–12 shows the PDE for a 3×3 array of a CF340 MPW2009 at 1.8 V excess bias voltage. Neighboring SPADs exhibit very similar photon detection efficiencies.

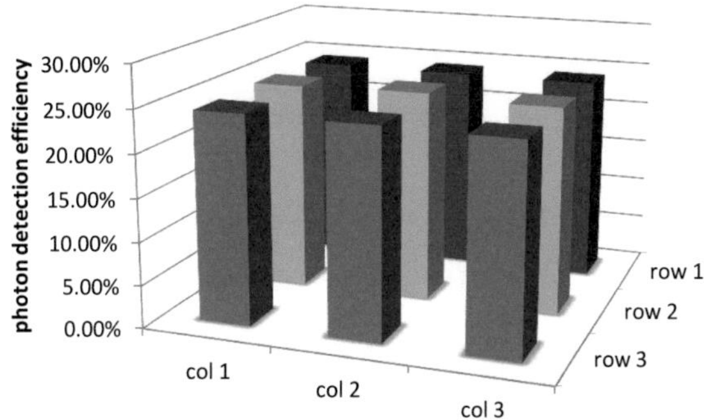

Fig. 4–12: Photon detection efficiency for 3×3 array in CF340 MPW2009 at 1.8 V excess bias voltage. All SPADs exhibit homogeneous detection efficiencies.

The variation over six samples is depicted in Fig. 4–13. Fixed voltages $V_{eb} = 1.8\,\text{V}$ and $|V_{bd}| = 13.3\,\text{V}$ are applied. For each sample, the span from minimum to maximum PDE value is shown. Mean values are connected with a solid line.

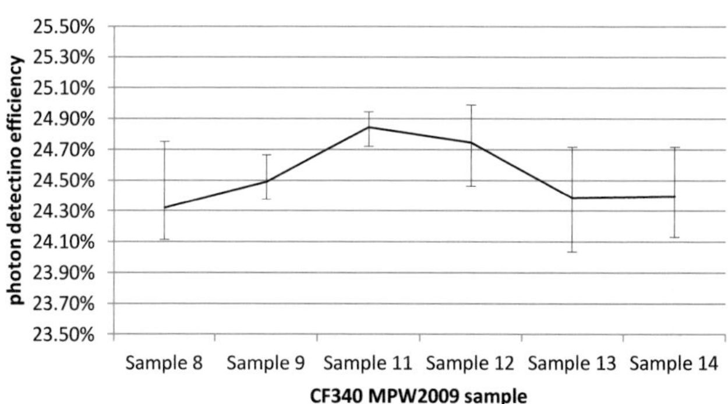

Fig. 4–13: Photon detection efficiency for multiple samples of 9 SPADs each. Mean values and span from minimum to maximum measured value are shown.

The signals of multiple illuminated SPADs can be combined for a higher overall SNR, as described in Section 3.4.3.4 where saturation effects are discussed. A homogeneous PDE over the entire array gives comparable count rates for all pixels (apart from dark counts) and does not require controlling the excess bias voltage for each pixel individually. One breakdown voltage and one excess bias voltage are sufficient for the entire array.

However, because each SPAD has a slightly different breakdown voltage, the effective excess bias voltage is different for each SPAD when a fixed voltage $V_{eb}+|V_{bd}|$ is applied. Devices which show a higher breakdown voltage therefore effectively operate at lower excess bias voltage. The expected behavior would be a lower PDE for devices with a higher breakdown voltage, thus a negative correlation coefficient $\rho_{V_{bd},\text{PDE}}$. This hypothesis is tested with the correlation coefficient between V_{bd} and PDE of the $n = 54$ SPADs (6 samples with 9 SPADs each). With the mean values $\mu_{V_{bd}}$, μ_{PDE} and the standard deviations $\sigma_{V_{bd}}$, σ_{PDE} of V_{bd} and PDE, the correlation coefficient is given by

$$\rho_{V_{bd},\text{PDE}} = \frac{\text{cov}(V_{bd},\text{PDE})}{\sigma_{V_{bd}}\sigma_{\text{PDE}}} = \frac{\frac{1}{n}\sum_{k=1}^{n}(V_{eb,k} - \mu_{V_{bd}})(\text{PDE}_k - \mu_{\text{PDE}})}{\sigma_{Vbd}\sigma_{\text{PDE}}}. \tag{4.1}$$

As expected, the correlation coefficient $\rho_{V_{bd},\text{PDE}} = -0.44$ is negative.

4.1.2.3 Count Rate

The count rate depends on the type of front-end circuit and associated dead time. Graphs of count rate vs. photon rate are shown in Section 4.1.3 where measurements for both passive and active quenching and recharge circuits are presented.

4.1.2.4 Dark Count Rate

A histogram of dark count rates (DCR, see Section 3.2.3.2) for a set of 54 SPADs is shown in Fig. 4–14. Most of the devices exhibit a DCR of some hundred Hertz. Besides these low-noise pixels, there are a few pixels in the hundred kHz range up to 260 kHz. Therefore the mean DCR does not represent the majority of SPADs. The median DCR is a more appropriate measure. The few noisy pixels are referred to as 'screamers'.

A possible explanation for the few high-DCR pixels are local defects in the semiconductor. Lattice defects or impurities give rise to generation-recombination-centers that increase the DCR. A comparison of multiple 3×3 arrays did not show specific patterns of similar DCR for the SPADs of different arrays. SPADs at random positions of the array suffer from a high DCR. A correlation with neighboring pixels or correlation between different chips is not observed. If several pixels are illuminated in a laser rangefinder, high DCR pixels that have detrimental effect on the SNR can be disabled. DCR increases with excess bias voltage as shown in Fig. 4–15.

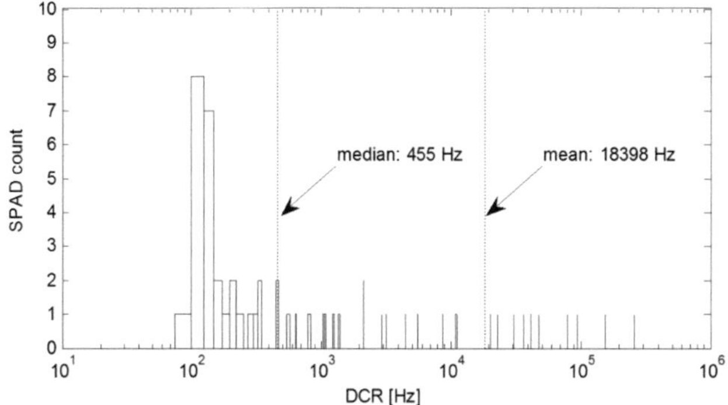

Fig. 4–14: Histogram of SPAD counts in darkness for 54 SPADs at V_{eb}=1.8 V, room temperature and 25 Hz bin width. The majority of pixels show a low dark count rate (DCR) of few 100 Hz, the median DCR is 455 Hz. Because of a few very noisy SPADs of up to 260 kHz the mean DCR of 18398 Hz is significantly higher than the median.

A low DCR is a major design goal for detectors in low-noise applications. A laser rangefinder typically operates in ambient light. As long as photon noise from ambient light is the dominant noise source, a certain level of DCR can be tolerated.

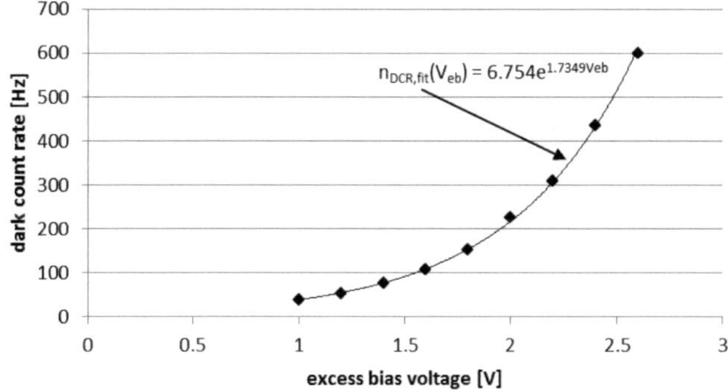

Fig. 4–15: Dark count rate over excess bias voltage. Measurement values with exponential fit.

4.1.2.5 Dead Time

The dead time (t_{dead}, see Section 3.2.3.3) of the SPAD is the time interval during which no photons can be detected. The inter-arrival time is the time difference between two subsequent SPAD pulses. The dead time can be determined from an inter-arrival time plot as shown in Fig. 4–16. The SPAD is illuminated with an LED as a light source that emits Poisson distributed photons (see Section 3.2.2). A real-time sampling scope records a trace of SPAD pulses. The histogram of the time differences between subsequent pules in Fig. 4–16 is generated from that trace. The minimum inter-arrival time between a first photon at time $t = 0$ and a subsequent photon gives the dead time $t_{dead} = 30\,\text{ns}$.

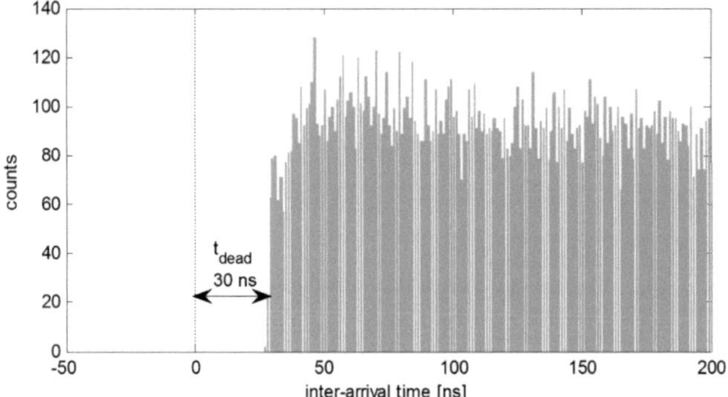

Fig. 4–16: Inter-arrival time plot for passively quenched SPAD at V_{eb}=1.8 V and V_{eb}-V_{quench}=0 V. Dead time (DT) is 30 ns.

In passive quenching circuits (see Section 3.4.1), the dead time can be adjusted via the quenching/recharge resistance. A dead time range from 160 ns down to 14 ns can be set in the SPADDEVEL test cell by controlling the bias voltage of the quenching transistor. The voltage difference between V_{eb} and V_{quench} defines the gate-source voltage of the transistor. Fig. 4–17 shows the adjustment range.

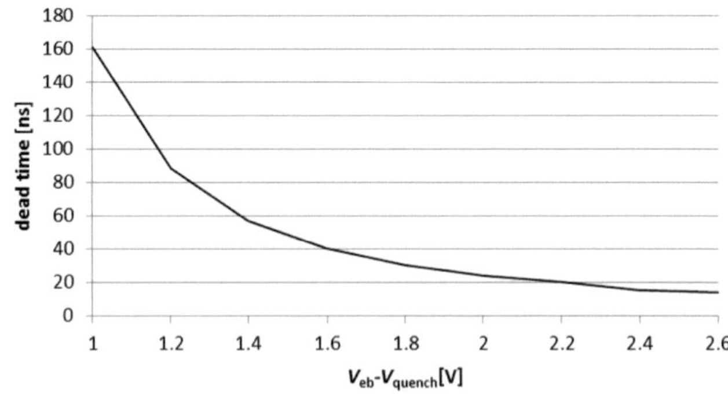

Fig. 4–17: SPAD dead time for different bias voltages of the quenching transistor.

In active quenching (see Section 3.4.2), the hold-off time can be arbitrarily chosen by an appropriate delay element which sets the hold-off time. The hold-off time in the present active quenching circuit ranges from 12 µs down to 5.4 ns.

4.1.2.6 Afterpulsing of a SPAD

Afterpulsing (see Section 3.2.3.5) has to be analyzed both in terms of magnitude and distribution over time. As discussed in Section 3.5.3, a laser rangefinder can tolerate high degree of afterpulsing if the afterpulsing is distributed over several RF periods.

Fig. 4–18 shows the conditional detection probability density of detecting an event during a 1 ns wide time window (bin) at a given time after an initial event at time zero. Directly after the initial event, during dead time, no events are detected.

Under the assumption that all counts are independent and Poisson distributed, each time window would show the same average number of counts. The counts occurring a sufficiently long time, for example five times the decay time, after the initial event can be assumed independent of the initial event. After that time, the (random) dark counts can be determined independent of the initial event at time zero.

At low count rates, some bins will count a pulse, others may not. A horizontal line marks the mean dark count value per bin in Fig. 4–18 for events that are not correlated with the initial event at time zero. On average each time window sees less than one count. The uncorrelated count rate is $n_{\text{random}} = 3 \cdot 10^{-7} / \text{ns} = 300\,\text{Hz}$.

Afterpulses are correlated with the initial pulse at time zero. They can be clearly identified from the high conditional probability densities on top of the dark counts. The measurement in Fig. 4–18 gives an afterpulse probability $\text{Prob}_{AP} = 1.28\%$ and a decay constant of 11 ns for an active quenching circuit operated at 5.4 ns dead time. For longer dead times, the afterpulse probability decreases because trapped carriers released during dead time do not trigger a SPAD pulse (see Table 4.1 on p. 112).

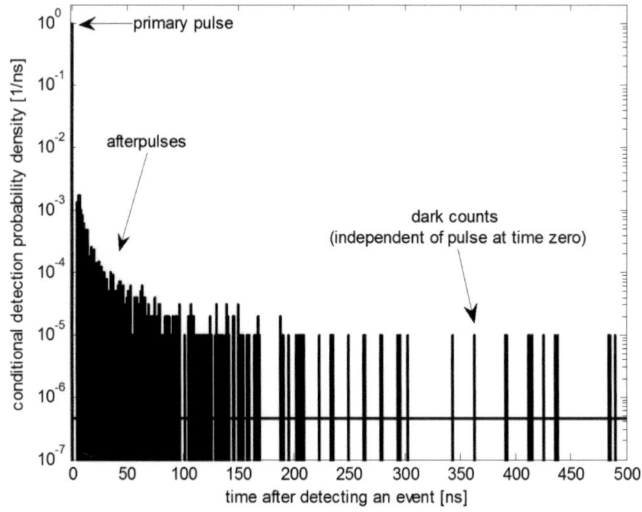

Fig. 4–18: Conditional detection probability density function (PDF) of a detection event, provided a photon has been detected at time 0. Integration over the PDF from 5 ns to 275 ns gives 1.28 % afterpulse probability in addition to the dark count probability.

4.1.2.7 Crosstalk between neighboring SPADs

Crosstalk is evaluated similar to afterpulsing by determining the conditional probability of detecting a pulse at a detector B, provided a detector A has detected a primary photon. Crosstalk is measured using the individual outputs of the 3×3 array from the SPADDEVEL test chip with 26 µm pixel pitch. Different combinations of pixels A and B are selected from the array. Even direct neighbors show negligible crosstalk barely exceeding the level of dark counts. Crosstalk is expected to decrease even further at larger pixel pitch. Negligible crosstalk can be inferred for the CF340 MPW2011 chips which have a larger pixel separation.

4.1.2.8 Jitter

The jitter measurement setup comprises an oscilloscope with high sampling rate (20 GSa/s, LeCroy Wavemaster 8600A), a short-pulse diode laser with laser driver (76 ps FWHM, Becker & Hickl BHL-600) and an attenuator. The laser module generates a train of short optical pulses and an electrical synchronization signal for each pulse. About one in a hundred optical pulses generates a SPAD output pulse in order to avoid pile-up (see Section 3.5.2). The oscilloscope precisely measures the time difference between synchronization signal of the laser and SPAD output pulse. Fig. 4–19 shows SPAD jitter as a histogram of these time differences. It should be noted that the histogram includes the laser pulse width ($\tau_{\text{laser}} = 76\,\text{ps FWHM}$), the synchronization jitter ($\tau_{\text{sync}} = 10\,\text{ps FWHM}$) and the oscilloscope jitter ($\tau_{\text{scope}} = 2.5\,\text{ps FWHM}$) which broaden the measured histogram. Assuming Gaussian error propagation and statistically independent random variables, these parameters can be subtracted [74]. The SPAD jitter $\tau_{\text{SPAD}} = 77\,\text{ps}$ can be calculated from the measured jitter $\tau_{\text{meas}} = 109\,\text{ps}$ FWHM,

$$\tau_{SPAD} = \sqrt{\tau_{\text{meas}}^{2} - \tau_{\text{laser}}^{2} - \tau_{\text{sync}}^{2} - \tau_{\text{scope}}^{2}}. \tag{4.2}$$

The jitter curve in Fig. 4–19 shows an asymmetric distribution. A possible explanation for this asymmetry is a diffusion tail (see Section 3.2.3.6) at the trailing edge.

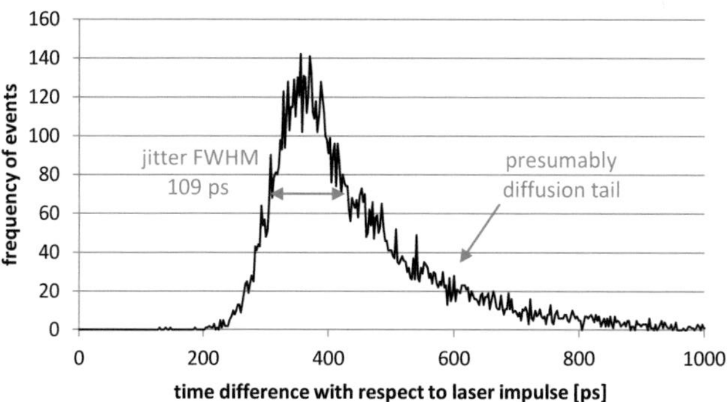

Fig. 4–19: SPAD jitter

An alternative jitter measurement setup employs a high-precision time-to-amplitude converter (TAC, Becker & Hickl SPC-130) instead of an oscilloscope to measure the time difference between synchronization signal and SPAD output pulse. The other components of the jitter measurement setup are the same. The TAC allows an even finer timing resolution (see [92] for more details on the TAC).

In the shown experiment, the LeCroy oscilloscope is used for jitter measurement because the oscilloscope can be easily co-integrated in an Matlab environment for automated measurements.

4.1.3 Active / Passive Quenching

The behavior of SPADs with active and passive quenching circuit (see Section 3.4) is analyzed with a sensitivity measurement which plots count rate over incident photon rate. An ideal detector has a linear relationship between count rate and photon rate. The slope is defined by the photon detection efficiency η_{PDE}. For any detector with dead time, the count rate is ultimately limited by the inverse of the dead time as an upper threshold (Eq. (3.18)).

4.1.3.1 Passive Quenching

The peak count rate of the passive quenching circuit is accurately predicted via the paralyzable detector model (see Section 3.4.3.2) by $n_{\mathrm{max}} = 1/(e t_{\mathrm{dead}})$. The measured count rate deviates from the paralyzable detector model at high photon rates (Fig. 4–20), but the hybrid detector model (see Section 3.4.3.3) nicely fits the measured count rates.

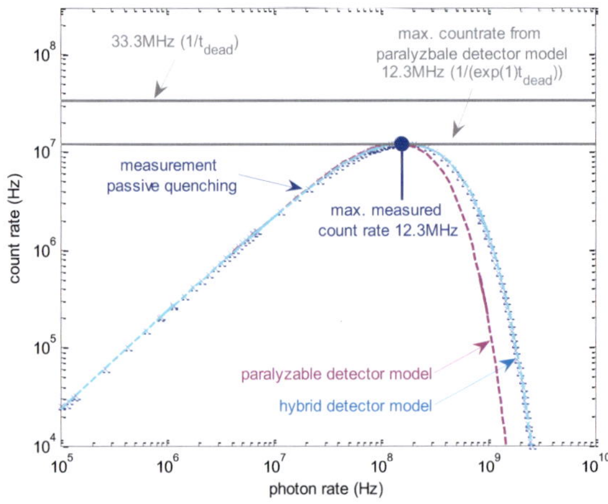

Fig. 4–20: Sensitivity measurements for fabricated passive circuit in comparison with a paralyzable [87] and hybrid [88] detector model. Dead time 30 ns, excess bias voltage 1.80 V. Hybrid model fit to measurement data with 21 ns non-paralyzable and 14 ns paralyzable dead time.

As the photon detection efficiency (PDE) recovers with increasing V_{eb}, the probability of triggering another avalanche shortly after avalanche quenching is reduced. This results in a lower effective photon rate seen by the detector. Hence in practice there are fewer events that paralyze the SPAD than in the theoretical detector models (see Section 3.4.3) that assume a constant PDE. This translates to an under-estimated count rate at high photon rates in the paralyzable detector model. The reduction of PDE is also indicated by the experimental inter-arrival time plot shown in Fig. 4–21. The inter-arrival time denotes the time interval between two subsequent events. Right after the dead-time the measured number of counts is not immediately at the theoretical exponential decay curve (red) but recovers towards the theoretical curve.

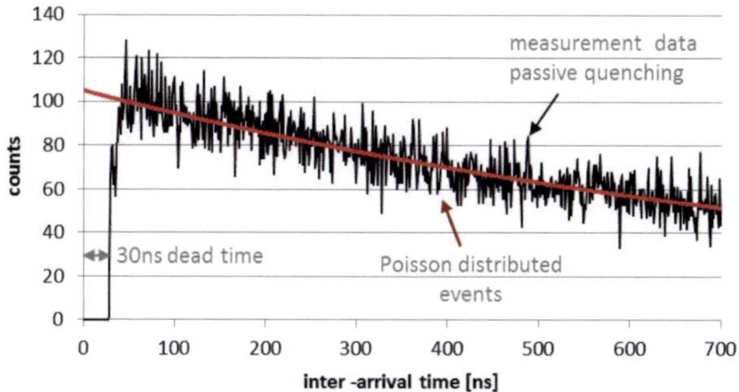

Fig. 4–21: Inter-arrival time histogram for passive circuit with 30 ns dead tine.

The exponentially decay curve of the inter-arrival time is expected from the Poisson distribution of events (see Section 3.2.2 on Photon Statistics and Appendix B). For accurate predictions, SPAD floating node voltage curves and PDE dependence on V_{eb} must be taken into account.

4.1.3.2 Active Quenching

Fig. 4–22 shows the count rate of an active quenching circuit over photon rate. The active quenching circuit actually reaches the theoretical limit of $1/t_{dead}$ and nicely matches the saturable detector model (see Section 3.4.3.1).

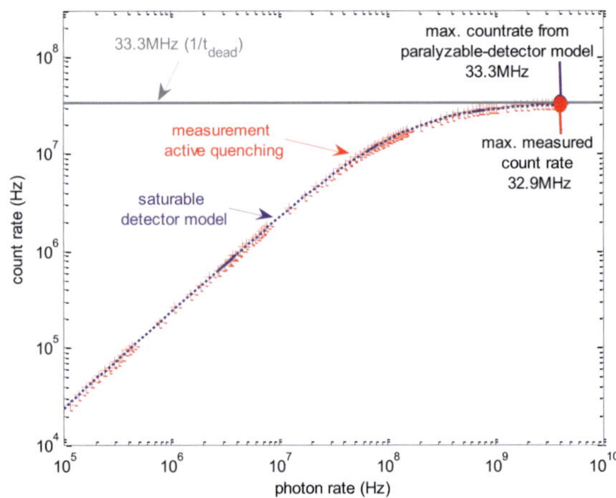

Fig. 4–22: Sensitivity measurements for fabricated active quenching circuit in comparison with saturable detector model [87].

Fig. 4–23 shows an inter-arrival time plot for the active quenching circuit. Other than in passive quenching, the inter-arrival-time plot of the analyzed active quenching circuit shows an initial peak with more than twice the expected counts. A possible explanation is that photons incident during strong active recharge trigger an avalanche that releases an output pulse right after dead time (see Section 3.5.4). A possible explanation for the oscillation after the peak is ringing of the circuit after the strong recharge current.

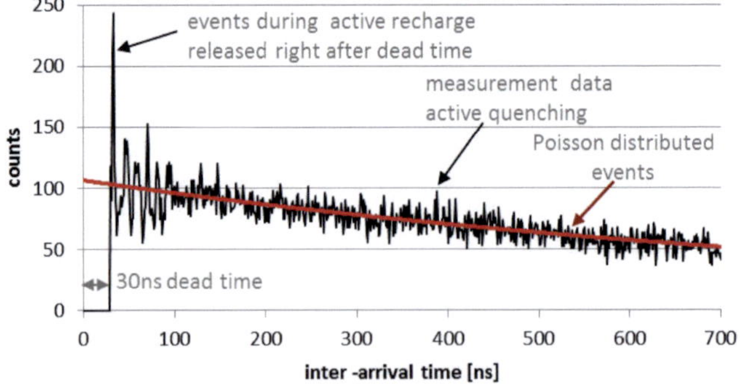

Fig. 4–23: Inter-arrival time histogram for implemented active circuit at 30 ns dead tine.

For a dead time of 5.4 ns and an excess bias voltage of 2.6 V, the actively quenched SPAD reaches a maximum count rate of 185 MHz, Fig. 4–24. The dynamic range as defined in Eq. (3.19) is 139 dB. A count rate of 185 MHz and a dynamic range of 139 dB represent, to the best of our knowledge, the highest count rate and dynamic range for single-photon avalanche diodes reported so far [53]. Comparing to state-of-the-art results [58], the dynamic range is improved by 6 dB, and the count rate is increased by a factor of 3.

A commercially available SPAD (idQuantique id101-20) is measured as a reference. It should be noted that this device shows paralyzable behavior despite its active quenching circuit. In this work a powerful recharge transistor M_3 (Fig. 3–18) prevents paralysis during recharge.

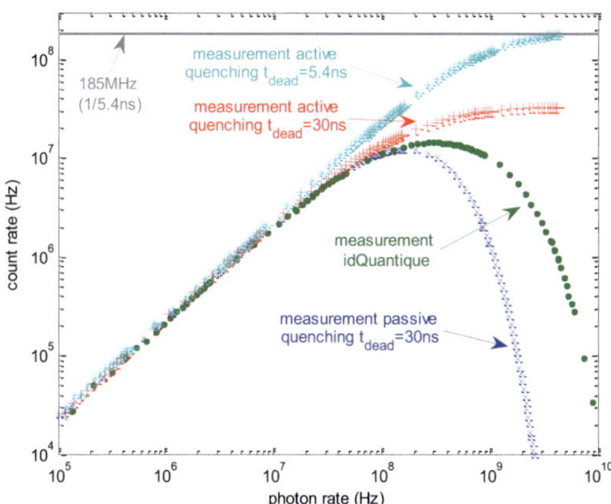

Fig. 4–24: Sensitivity measurements for a self-fabricated passive quenching circuit, a commercially available active quenching SPAD (idQuantique id101-20), and a self-fabricated active quenching circuit. At 5.4 ns dead time, the actively quenched SPAD is operated at 2.6 V excess bias voltage.

Table 4.1 concludes this section with performance parameters for passively and actively quenched SPADs. The actively quenched SPAD is measured twice at 1.8 V and 2.6 V excess bias voltage because the active quenching circuit required a higher operating voltage to achieve a short dead time. A commercially available device from idQuantique is listed as a reference.

Device	Excess bias voltage V_{eb}	Dead Time t_{dead}	Dark Count Rate[*] (n_{DCR})	Poisson RMS DCR[*] ($\sqrt{n_{\text{DCR}}}$)	Max. Count Rate (n_{max})	Dynamic Range[**]	APP ($\pm 1\,\sigma$)
	[V]	[ns]	[Hz]	[Hz$^{1/2}$]	[MHz]	[dB]	[%]
idQuantique (id101-20)	N/A	39	374.0	19.3	14.4	117.4	0.523 (± 0.135)
Passive	1.8	30	153.0	12.4	12.3	119.9	0.567 (± 0.076)
Active	1.8	30	110.0	10.5	32.9	129.9	0.428 (± 0.064)
Active	2.6	5.4	410.8	20.3	185.0	139.2	1.28 (± 0.036)

[*] at room temperature [**] Dynamic range $\mathrm{DR} = 20\log\left(\dfrac{n_{\text{max}}}{\sqrt{n_{\text{DCR}}}}\sqrt{t_{\text{meas}}}\right)$; $t_{\text{meas}} = 1\,\mathrm{s}$

Table 4.1: Performance parameters for SPADs with passive and active quenching circuits

4.2 Experimental Results MPW2011

New test structures are specified based on the results from the previous test chips. Design and fabrication is done by STMicroelectronics in cooperation with Bosch Automotive Electronics. The experiments presented in this section refer to either new devices or measurements that have not been performed in detail with previous test chips as discussed in Section 4.1.

4.2.1 Test Structures

There are six different SPAD variants implemented on the multi-project wafer run 2011 (MPW2011). The 8 µm diameter disk-shaped SPADs from the previous wafer (MPW2009) are implemented for reference alongside with more area efficient SPADs that are square with rounded corners. For a higher fill factor in a SPAD array, an 'aggressive' variant of the square SPAD with rounded corners is implemented that has a larger active area but reduced guard structures (see e.g. Fig. 3–12). Each SPAD variant is implemented in positive- and negative-drive option (see Section 3.4.4). SPAD shapes have previously been depicted in Fig. 4–2. Table 4.2 lists the diameters and the active areas for the implemented SPAD variants.

The shape of the square SPADs with rounded corners is a superellipse (or Lamé curve) [93]. Rounded corners are implemented to prevent electric field peaks and premature breakdown in the corners [64, 94].

Shape	Active area (base diameter, area)	
	Negative-drive (ND)	Positive-drive (PD)
Disk (type 1)	8 µm ø 50.3 µm²	8 µm ø 50.3 µm²
Superellipse (type 2)	8 µm ø 63.1 µm²	8 µm ø 63.1 µm²
Superellipse 'aggressive' (type 3)	10.5 µm ø 101.5 µm²	9 µm ø 75.1 µm²

Table 4.2: SPAD variants in MPW2011 test chips

An array of SPADs comprises individual pixels or cells. Each pixel incorporates front-end circuitry for quenching, control logic, lines for supply voltages and routing of signals to and from the SPAD. Fig. 4–25 shows the layout of a section of the SPAD cell with output buffer, quenching, control logic and supply lines. The objective in cell design is a high fill factor (see Section 3.2.3.8).

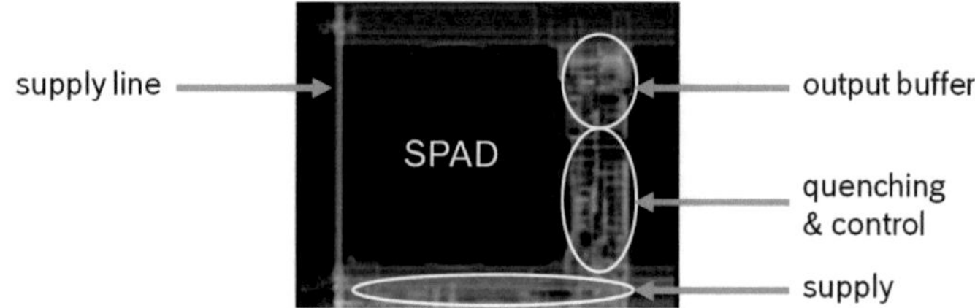

Fig. 4–25: Layout of a SPAD cell

There are two test chips implemented on the multi-project wafer 2011. Chip 1 (MBSAT) consists of various 2x3 SPAD arrays with access to individual SPAD outputs. Front-end circuitry for both active and passive quenching is available. Besides the CMOS inverter with fixed threshold for detecting SPAD pulses, an additional operational amplifier with one input connected to the SPAD floating node voltage and one input connected to a reference voltage, shown in Fig. 4–27, is added to track the SPAD floating node voltage during experiments (see Section 4.2.2.1).

Chip 2 (CF340 MPW2011, ASIC 2) includes an array of 6×10 SPADs. The array is segmented into two halves to two test different SPAD variants in one array. There is no simultaneous access to individual SPAD outputs. The CF340 MPW2011, designed by Bosch Automotive Electronics, also contains means for time-of-flight distance measurement that will be described in Chapters 5 and 6. Fig. 4–26 shows photomicrographs of the test chip in a CQFP-package.

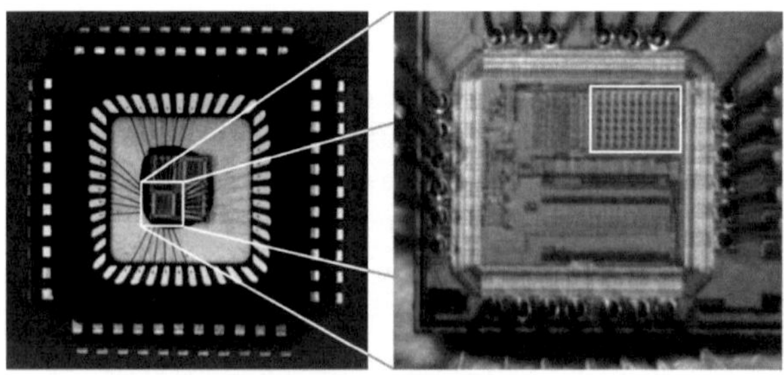

Fig. 4–26: Photomicrograph and blow-up of CF340 MPW2011 test chip
with 6×10 SPAD array in top right corner

4.2.2 SPAD Cell Performance Parameters

The characterization of the test chips is split into SPAD cell and SPAD array performance parameters. MBSAT test chips are used where direct access to the cell output is required. Further parameters are measured more conveniently in the CF340 MPW2011 test chips, which include a counter and timing circuitry. Unless otherwise stated, typical measurements are given for the type 3 SPADs (see Table 4.2).

4.2.2.1 SPAD Floating Node Voltage

An additional operational amplifier provides a convenient method to measure the SPAD floating node voltage V_S. Fig. 4–27 sketches the circuit for a passively quenched SPAD. When V_S drops below the decision threshold voltage of the CMOS inverter, a SPAD pulse is detected at the output $V_{\text{OUT, inverter}}$. For the CMOS inverter the decision threshold voltage is fixed at $V_{DD}/2$. When V_S drops below the reference voltage V_{ref} of the operational amplifier, a SPAD pulse is detected at the output $V_{\text{OUT,OpAmp}}$.

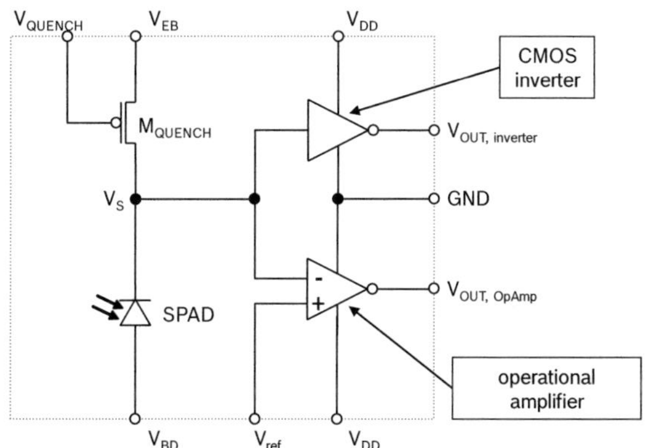

Fig. 4–27: SPAD cell with additional variable threshold comparator

The SPAD floating node voltage as a function of time is measured as follows. The SPAD is illuminated at a constant light level and generates a series of output pulses. The outputs of the CMOS inverter and of the operational amplifier are measured with two channels of an oscilloscope. The time difference between the raising edge of a SPAD pulse at $V_{\text{OUT,inverter}}$ and the raising edge of the SPAD pulse at $V_{\text{OUT,OpAmp}}$ as well as the time difference between the raising edge of the SPAD pulse at $V_{\text{OUT,inverter}}$ and the falling edge of the same SPAD pulse at $V_{\text{OUT,OpAmp}}$ are determined from the oscilloscope traces of the two channels. This measurement is repeated for different levels of the reference voltage V_{ref} of the operational amplifier. This way, the transient behavior of the SPAD moving node voltage V_S can be measured.

Fig. 4–28 shows three measurement curves for an 8 μm diameter disk-shaped ND-SPAD, one curve for passive quenching (and passive recharge) and two for active quenching (and active recharge) with different hold-off time. In this measurement the excess bias voltage is $V_{eb} = 1.8\,\text{V}$. The reference voltages of the operational amplifier is $V_{\text{ref}} = 0.2...1.6\,\text{V}$, which almost spans the range from 10 % V_{eb} (0.18 V) to 90 % V_{eb} (1.62 V). The quantity $V_{th} = 50\%\,V_{eb} = 0.9\,\text{V}$ denotes the decision threshold voltage of the CMOS inverter. The dead time t_{dead} is measured as the time difference between the time when V_S drops below V_{th} and the time when V_S raises above V_{th}.

Fig. 4–28 clearly shows the difference between passive and active quenching and recharge. With passive recharge, V_S recovers slowly to V_{eb}. The recharge time to recover from 10% to 90% V_{eb} is about 60 ns. With active recharge, V_S quickly recovers to V_{eb} after a fixed hold-off time. The recharge time to recover from 10% to 90% V_{eb} is less than 3 ns. This behavior corresponds to the theoretical voltage curves sketched in Fig. 3–16 (b) and Fig. 3–17 (b). Active recharge is especially well-suited for larger area SPADs because the powerful active recharge transistor quickly recharges the corresponding higher capacitance of a large device.

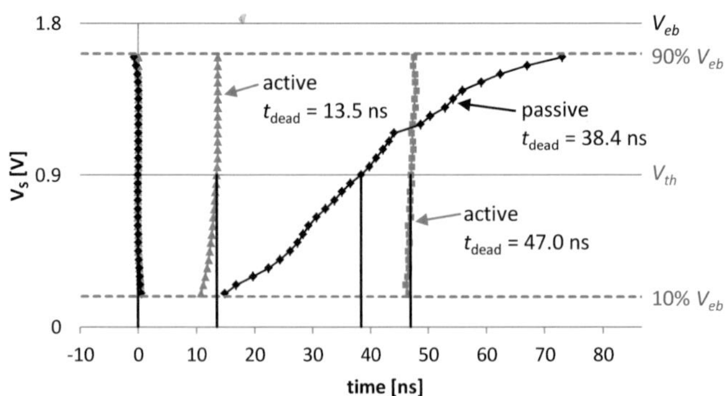

Fig. 4–28: SPAD floating node voltage V_S as a function of time. Disk-shaped SPAD with 8 μm diameter in negative-drive option; one curve for passive quenching and recharge; two curves for active quenching and recharge with different hold-off times.

4.2.2.2 Jitter

The same jitter measurement setup is used for the MBSAT test chips as previously in Section 4.1.2.8. Background information on jitter is provided in Section 3.2.3.6. Fig. 4–29 shows jitter measurements for different SPAD types in negative-drive option (see Table 4.2). The measured jitter FWHM (full width at half maximum) is 120 ps for the type 1 SPAD, 132 ps for the type 2 SPAD and 132 ps for the type 3 SPAD. A jitter measurements at different temperatures show limited influence of the temperature in the order of some picoseconds.

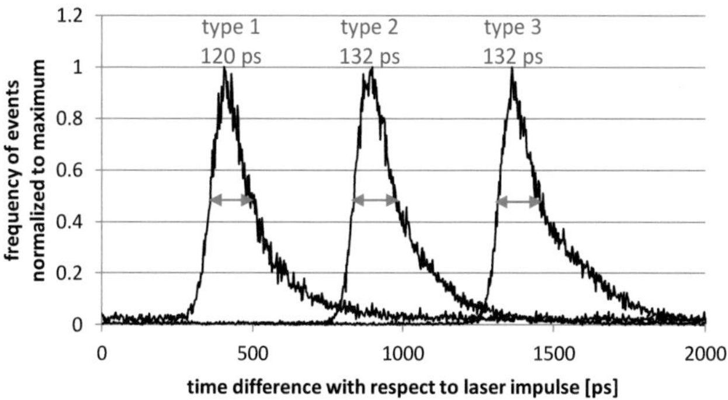

Fig. 4–29: Jitter for different SPAD types (see Table 4.2) in negative-drive option. The histograms are normalized to a maximum value of 1. Measurements are labeled with jitter values (full width at half maximum).

4.2.3 SPAD Array Performance Parameters

4.2.3.1 Photon Detection Efficiency

Fig. 4–30 depicts the photon detection efficiency (PDE) of a 6×10 SPAD array with type 3 ND-SPADs. All pixels show comparable PDE with a mean value $\mu = 24.4\%$ and a standard deviation $\sigma = 0.22\%$. The devices are operated at an excess bias voltage of $V_{eb} = 1.8\,\mathrm{V}$ and an applied breakdown voltage of $V_{bd} = -13.3\,\mathrm{V}$. Light from an integrating sphere with internal red LED centered at 638 nm wavelength homogeneously illuminates the SPAD array. Prior to illumination, the dark count rate is measured and then subtracted from the count rate measured under illumination.

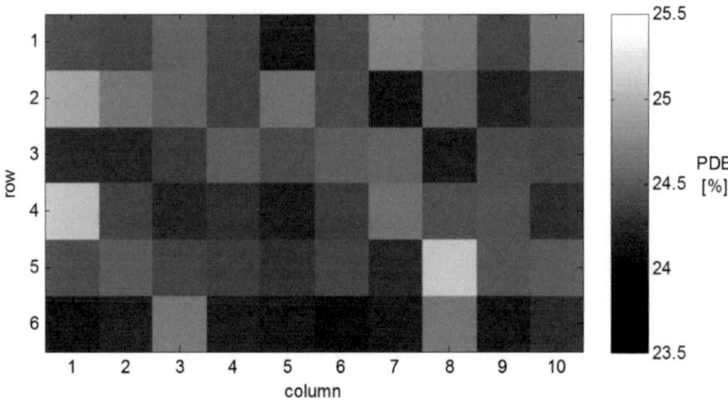

Fig. 4–30: Photon detection efficiency (PDE) for 6×10 array in CF340 MPW2011 at excess bias voltage of 1.8 V. The PDE is gray scale coded. The standard deviation of the PDE for this chip is σ = 0.22 %, the mean value μ = 24.4 %.

The PDE variation for four samples is depicted in Fig. 4–31. For each chip the span from minimum to maximum value is shown. Mean values are connected with a solid line.

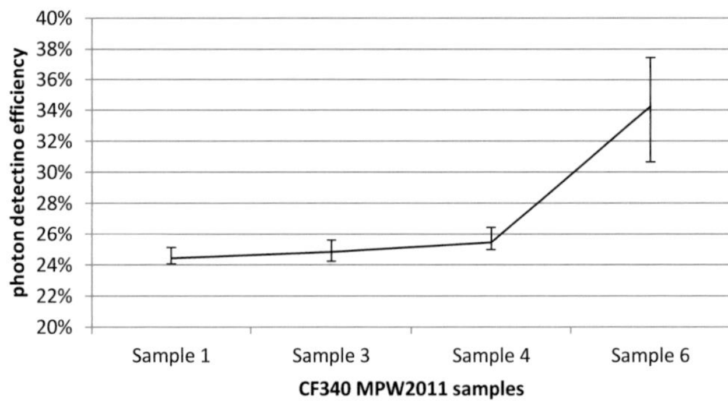

Fig. 4–31: Photon detection efficiency for multiple chips of 60 SPADs each. Mean values and the span from minimum to maximum measured value are shown. Because of high afterpulsing, sample 6 seemingly shows a higher PDE.

Samples 1, 3, 4 range around 25 % whereas sample 6 seemingly excels with 34 % PDE. Also It should be noted that the PDE of sample 6 spans a wider range from minimum to maximum value. As will be shown in the next two sections, sample 6 suffers from high dark count rate and high afterpulse probability. Dark counts are subtracted. However, afterpulses give rise to on average more than one output pulse per detected photon, which seemingly increases the PDE.

4.2.3.2 Dark Count Rate

A low average dark count rate (DCR, see 3.2.3.2) of several hundred Hz has been measured with MPW2009 test chips and was also expected for the MPW2011. A modification of the manufacturing process in the semiconductor fabrication plant, however, accidentally increased the number of traps in the MPW2011. Fig. 4–32 shows the percentage of SPADs below a given DCR. As a worst-case example board 6 suffers from a median DCR of 338 kHz (mean 402 kHz). Other devices from the same wafer have several 10 kHz median DCR with few noisy SPADs (screamers).

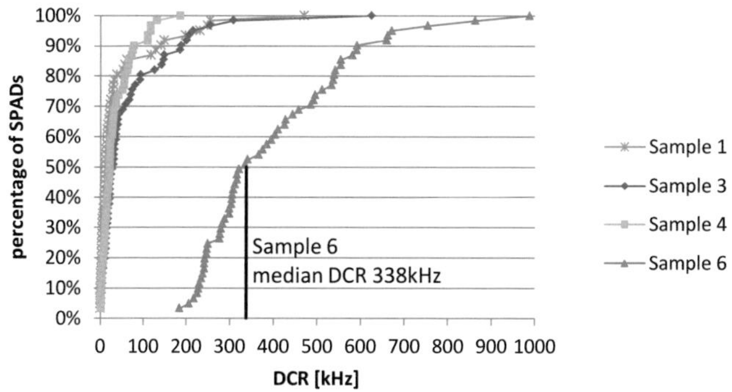

Fig. 4–32: Dark count rate (DCR) for four samples with 6×10 SPAD arrays. The percentage of SPADs on the vertical axis denotes the cumulative percentage of SPADs below a given DCR.

4.2.3.3 Afterpulsing

A higher trap concentration, indicated by the high DCR, also gives rise to high afterpulsing probability. All measurements of the afterpulse probability are done at room temperature. A SPAD from chip 3 with a DCR of 21 kHz has an afterpulse probability of 2.0 %. Fig. 4–33 shows an inter-arrival time plot (see Section 3.2.2) for the extreme case of a SPAD from sample 6. The measured occurrence of short inter-arrival times deviates significantly for the exponential decay that is expected from Poisson distributed counts.

Sample 6 shows a DCR of 425 kHz and a high afterpulse probability of 28 %, despite a long dead time of 48 ns. As explained in Section 3.2.3.5, a long dead time is expected to reduce afterpulsing because carriers that are released during the dead time propagate through the (still reverse biased) junction without triggering a self-sustaining avalanche.

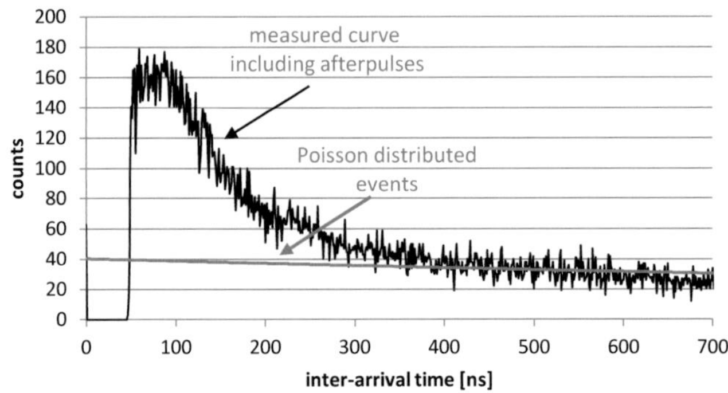

Fig. 4–33: Inter-arrival time plot of SPAD with a DCR of 425 kHz
and an afterpulse probability of 28 % despite a dead time of 48 ns.

For comparison, the afterpulse measurement shown in Section 4.1.2.6 of a
sample from the previous wafer run showed an afterpulse probability of 1.28%
at a short dead time of 5.4 ns. However, low the DCR of few hundred Hz of the
previous test chips indicate that they do not suffer from a high trap
concentration.

However, not only the afterpulse probability, measured in a fixed time interval
after the initial pulse, but also the afterpulse distribution over time is important
for a laser rangefinder (see Section 3.5.3). The graph above shows afterpulses
distributed over more than 150 ns. Because this is long with respect to the RF
modulation period of 1 ns at a modulation frequency of 1 GHz these afterpulses
are spread over several periods and will not directly cause distance errors. The
afterpulses are rather seen as a higher 'background light level' that reduces the
signal-to-noise ratio in a laser rangefinder.

4.2.4 SPAD Drive Circuit Options

In positive-drive (PD) option, only the SPAD capacitance C_S has to be
discharged and recharged upon photon arrival (see Section 3.4.4, Fig. 3–25) but
not the parasitic capacitance C_p. In negative-drive (ND) option also the parasitic
deep-n-well / p-substrate capacitance C_p is connected to the SPAD floating node
and has to be discharged and recharged upon photon arrival. As a matter of fact,
C_p has a higher capacitance than the SPAD p-well / deep-n-well capacitance C_S
itself. A significantly lower current i_S into the V_{HV} pin is expected and plotted
versus the count rate for PD-SPADs, Fig. 4–34.

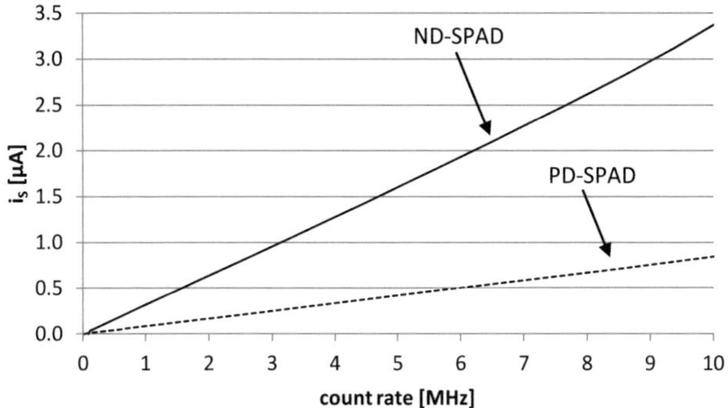

Fig. 4–34: Current i_S vs. count rate for a passively quenched type 1 SPAD in positive-drive (PD) and negative-drive (ND) option. The PD-SPAD shows a 3.8 times lower slope.

The count rate n is adjusted by sweeping the current i_{LED} of an LED light source. The quantity i_S denotes the mean current into the V_{HV} pin at a given count rate. Fortunately, the dark current is as small as 34 nA and can therefore be neglected compared to the average currents of i_S in the µA-range. The slope in Fig. 4–34 in A/Hz gives the average charge Q_{pp} in Coulomb per output pulse. This value gives the number of electrons per photon (strictly speaking per photon triggering an output pulse). The value depends on the capacitance and external circuitry.

$$Q_{pp} = \frac{\#\,\text{electrons}}{\text{SPAD pulse}} = \frac{i_S}{n}. \tag{4.3}$$

The measurement in Fig. 4–34 gives 2.0×10^6 electrons per pulse in negative-drive option compared to 0.53×10^6 electrons per pulse in positive-drive option. From the charge per pulse Q_{pp} and the operating voltage V, the total SPAD capacitance C can be calculated,

$$C = Q_{pp}V. \tag{4.4}$$

At an excess bias voltage $V_{eb} = 1.8$ V as the operating voltage V, the calculated capacitance is 182 fF in negative-drive and 47 fF in positive-drive option.

Fig. 4–35 shows a measurement of the count rate versus the LED current for both ND- and PD-SPADs at same bias conditions, equal quenching resistance and same active area. The by a factor 3.8 lower capacitance in PD option directly translates to a 3.8 times higher count rate because of fast discharge and recharge of the PD-SPAD.

Measurements with a PD-SPAD where the quenching resistance is further reduced yields a maximum count rate of 87 MHz. To the best of our knowledge this is the highest count rate achieved for a passively quenched SPAD.

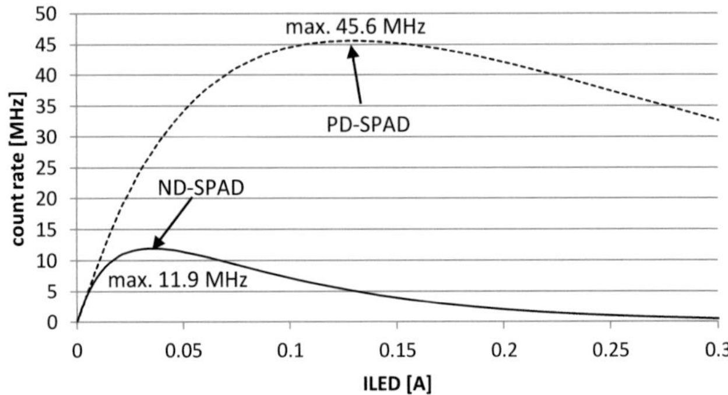

Fig. 4–35: Count rate vs. LED current ILED for a passively quenched type 1 SPAD in positive-drive (PD) and negative-drive (ND) option. At same quenching resistance, the PD-SPAD shows a 3.8 times higher maximum count rate n_{max}.

4.3 Conclusion

Together with the theoretical background in Chapter 3, the experimental results in this chapter help in understanding single-photon avalanche diodes. SPADs are promising candidates as detectors for laser rangefinders; however, care has to be taken with the optical system and front-end circuitry to operate within the limitations outlined in this chapter.

For a time-of-flight laser rangefinder, critical detector parameters either directly influence SPAD timing, or impact the optical and electrical system.

- Fill factor of SPAD array

  A high fill factor ensures efficient use of the incident light and allows using smaller (cheaper) receiver optics. Smaller features sizes in 90nm semiconductor fabrication processes and below will further increase the fill factor in the future.

- Photon detection efficiency

  Together with the fill factor, the PDE determines the light-gathering power of the optical system for achieving a desired count rate. The PDE can be adjusted in-chip via the excess bias voltage.

- Saturation behavior and quenching circuit

  The saturation behavior of a SPAD depends largely on the quenching circuit. When operating close to saturation, signal photons might be lost and the modulation efficiency decreases. Distance errors are expected because the timing behavior changes when (partial) recharge effects come into play. Band-pass filters can reduce overall irradiance to prevent saturation from background light while maintaining a high signal level.

- Supply voltage stability

  Stable supply voltages are required for accurate timing and defined detection efficiency. Possibly necessary external components for supply voltage generation, especially for biasing the SPAD above the breakdown voltage of 13-15 V, could diminish the expected cost benefit of a laser rangefinder ASIC.

- Array timing

  Timing varies from SPAD to SPAD in an array. Either a minimum timing difference can be ensured or a calibration is required.

Fortunately, a laser rangefinder poses relaxed requirements on some SPAD parameters.

- Dark count rate

  A laser rangefinder typically operates under ambient light. While 'screamers', i.e. SPADs with a very high dark count rate (DCR), may be detrimental in low light level applications, the effect in a laser rangefinder is the same as an increased background light level. This can be tolerated up to a certain extent and is investigated further in Chapter 5.

- Jitter

  Distance measurements integrate over a large number of events. Thus the average photon arrival time is evaluated rather than the arrival time of an individual photon. Nonetheless, the low-pass filter characteristic poses limits on tolerable jitter (more in Section 5.3.2.2).

- Afterpulsing

  Both, the afterpulse probability and the afterpulse distribution over time, determine the impact of afterpulsing on distance measurements. A laser rangefinder can tolerate a high degree of afterpulsing if the afterpulsing is distributed over several RF modulation periods.

From a business perspective the feasibility of a SPAD-based laser rangefinder largely depends on silicon area pricing. Moreover, the high-voltage capability and internal generation of stable supply voltages are crucial to avoid costly discrete external components. An on-chip bias voltage generation has been successfully tested in [95] for a 3×3 SPAD array, however, the feasibility of an on-chip bias voltage generation for larger arrays and at low cost still has to be evaluated.

A summary of the SPAD parameters in Table 4.3 concludes the detector analysis.

Performance Parameter*	MPW2009	MPW2011 (type 3 ND-SPAD)
Technology	130 nm CMOS imaging process	130 nm CMOS imaging process
Breakdown voltage	13.3 V	13.3 V
Typical excess bias voltage	1.8 V	1.8 V
SPAD drive circuit option	ND	ND and PD
Typical SPAD base diameter	8 μm	10.5 μm
Typical active area	50.3 μm	101.5 μm
Photon detection efficiency	23.5 %	24.4 %
Dark Count Rate (median)	455 Hz (from 54 SPADs)	32.5 kHz (from 240 type 3 SPADs)
Typical dead time	Passive: 30 ns Active: down to 5.4 ns	Passive: 48 ns Active: down to 10 ns
Afterpulse probability	< 1.5 %	2 − 28 %
Crosstalk probability	Negligible	Negligible
System jitter	109 ps	132 ps
Array size	3×3	6×10
Dynamic range	139 dB**	N/A

* all measurements at room temperature;

** measured with active quenching, 5.4ns at dead time

Table 4.3: Performance parameters for implemented SPADs

5 Laser Rangefinder System

5.1 Introduction

The key questions for a laser rangefinder are how far a distance can be measured under which ambient lighting conditions at what accuracy. The target of this chapter is a performance estimation of a SPAD-based laser rangefinder to answer these questions.

While sharing the basic optical properties with a conventional APD-based laser rangefinder in terms of optical power budget and lens system, the proposed SPAD-based laser rangefinder offers some peculiar features. A detector array is proposed [21, 22] instead of a single APD and an integrated on-chip reference detector becomes feasible at low cost [20]. The proposed SPAD-based laser rangefinder has a single application-specific integrated circuit (ASIC) that also comprises the photodetectors. SPAD output pulses can be directly processed in the digital domain without any additional analog filters or analog-to-digital converters. On the downside the rangefinder suffers from additional crosstalk between target and reference path inside the ASIC.

The proposed laser rangefinder measures the target distance by frequency domain-reflectometry (see Section 2.2.3) and more precisely using the single-photon synchronous detection concept (SPSD, see Section 2.3.3.2) that has first been outlined in [38]. In order to reach millimeter accuracy, the modulation frequency is increased to the GHz range. This RF requirement conflicts with the SPAD semiconductor manufacturing process which is optimized for good optical properties, i.e., thin layer spacing for good optical transmission but correspondingly high parasitic capacitances (see device structure in Section 3.3.3). This impacts electrical SPAD output pulses propagating over the chip, and severely disturbs the timing circuit (binning) that sorts SPAD pulses into sampling windows (bins) in synchronism with the laser modulator. A novel patent-pending bin homogenization scheme [96, 97] is presented in Section 5.3.5. The proposed bin homogenization scheme mitigates the problem of unequal bin widths and significantly improves distance accuracy.

The system performance including error contributions is modeled as sketched in Fig. 5–1. In a first step, labeled 'geometric model' the irradiance incident on the detector plane and geometric spot size over target distance are determined. An 'optical analysis' with detector array dimensions outputs the signal and background light power, modulation depth and SPAD count rates over target distance for further processing.

Error contributions are separated into contributions that can be described either analytically, by a Monte Carlo simulation, or that have to be estimated. The analytical description includes SPAD properties from Chapter 4 such as the photon detection efficiency, results from the optical analysis and both electrical and optical crosstalk. The binning process with and without bin homogenization is simulated with a Monte Carlo simulation. Delay drifts, essentially a variation of propagation delays with temperature, are based simulations and estimates by Bosch Automotive Electronics.

The performance estimation in Fig. 5–1 eventually only targets at a single measure: the distance error for a SPAD-based laser rangefinder that can be reached for a certain target distance and ambient light level. The results are presented in Section 5.4.

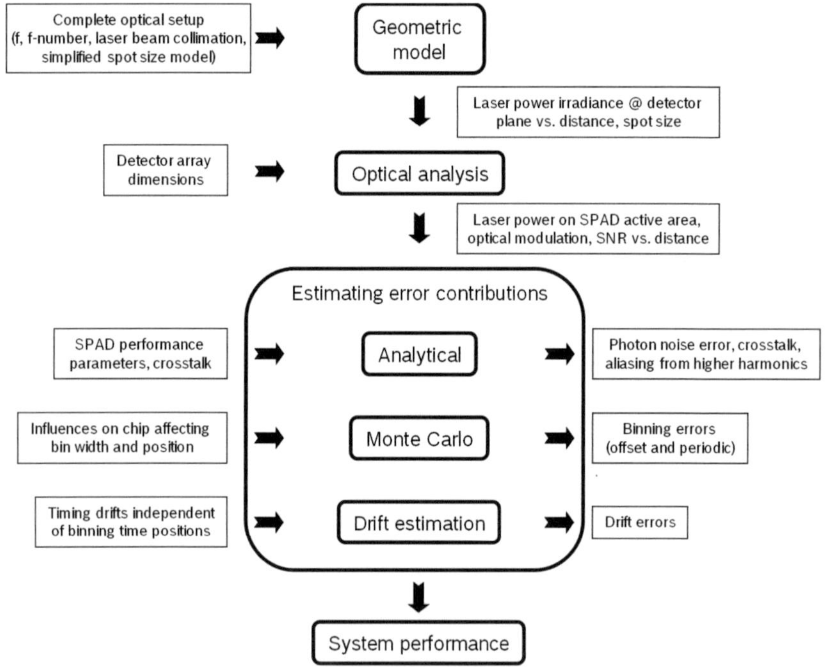

Fig. 5–1: Performance estimation concept

The target distance is calculated from the phase difference between sent and received signal. The error is averaged over the measurement time of a single-shot distance measurement. The influence of errors can therefore be visualized with a phasor diagram (Fig. 5–2). Unless otherwise stated, a worst-case impact of all error contributions, i.e., when the error phasor $\vec{E}$ is orthogonal to the measured phasor $\vec{M}$, is modeled. The distance error is calculated from the phase difference $\varphi_E = \varphi_M - \varphi_S$ between the actually measured phasor $\vec{M}$ and the desired signal phasor $\vec{S}$.

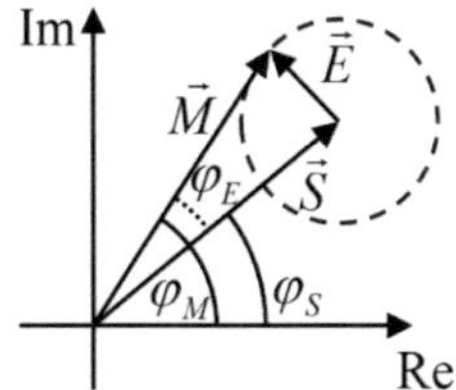

Fig. 5–2: Phasor diagram. $\vec{S}$ is the desired signal phasor, $\vec{E}$ is the error phasor, $\vec{M}$ is the resulting measured phasor.

5.2 Optics

The requirements for the optical system of a laser rangefinder differ considerably from an imaging camera. A camera actually projects a sharp image of an object on the detector plane. The optical system of an imaging camera typically features an auto-focus lens system to focus light from the scenery onto the film or image sensor, and a variable aperture and shutter to control the amount of light passing the lens system. These elements considerably add to system cost.

In a laser rangefinder we are not interested in a crisp and clear image of the target. The requirement is merely that sufficient signal light from the target strikes the detector active area. From a business perspective, low-cost components and simple manufacturing are required. The solution is a non-imaging fixed focus lens system with fixed aperture.

Fig. 5–3 sketches the optical system of a laser rangefinder. A collimator collects the light emitted from a laser diode and sends out a collimated beam onto the target. The size of the laser spot on the target depends on the divergence angle α and the target distance L. The target is modeled as an infinite plain surface that is an ideal Lambertian scatterer with reflectivity R. The receiver is characterized by the receiver lens diameter D, the focal length f and the actual position s' of the SPAD array (detector plane) with respect to the lens. The difference between the focal length and the position of the detector plane is given by $\delta f = s' - f$. An optical band-pass filter centered at the laser diode wavelength limits the amount of background light. The separation s_{parallax} of the optical axis of sending and receiving path defines the parallax and the (distance dependent) parallax angle θ in-between.

The following sections go through this optical system step-by-step to determine the intensity on the detector plane.

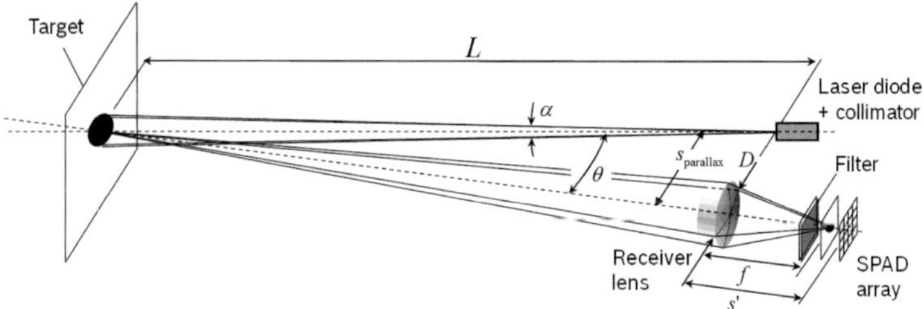

Fig. 5–3: Laser rangefinder optics. Fixed focus optical system with parallax between transmitting laser and receiver lens. Target distance L, laser beam divergence angle α, separation s_{parallax} of sending and receiving path, parallax angle θ, receiver lens diameter D, focal length f, actual position of detector plane with respect to receiver lens s'.

5.2.1 Background Light

Distances are to be measured at various ambient light scenarios. Table 5.1 gives an overview of different illumination levels. The target low-cost system aims at indoor-applications up to 10,000 lx.

Example	*Illuminance*
Direct sunlight	100,000 lx
Shade / overcast sky [*]	10,000 lx
Operating theater (surgery)	10,000 lx
TV studio	1,000 lx
Office lighting	500 lx
Street lights	10 lx
1 candela at 1m distance	1 lx
Starlight (new moon)	0.001 lx

* own measurement, other data from [6]

Table 5.1: Examples of illumination levels

The illuminance H is a photometric quantity that takes the wavelength-dependent sensitivity of the human eye into account. To draw conclusions about the photon flux, the illuminance in $[\text{lm/m}^2]$ has to be converted to its radiometric counterpart irradiance or intensity in $[\text{W/m}^2]$. Moreover, the spectral distribution of the light source has to be considered. Fig. 5–4 shows the spectral irradiance density of (a) a black body at 5900 K, (b) extraterrestrial sunlight and (c) terrestrial sunlight after having propagated through the earth's atmosphere. The atmosphere filters out a portion of the extraterrestrial sunlight. Oxygen, water and carbon-dioxide absorption are indicated.

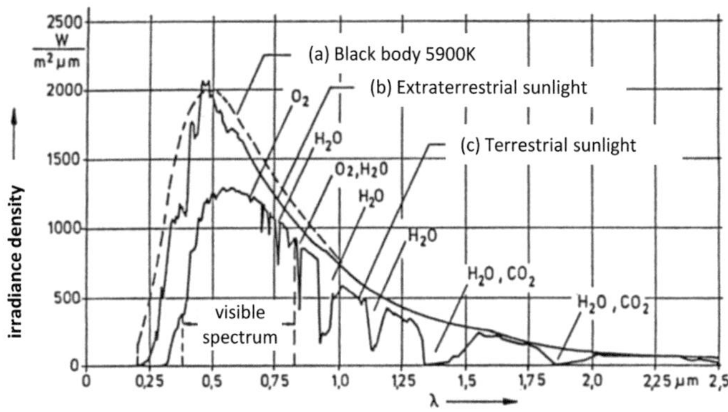

Fig. 5–4: Spectral background light irradiance density [98]

The spectral irradiance density of a blackbody radiator $I'(\lambda)$ is described with Planck's law in Eq. (5.1) [99]. It gives the energy per unit time that is emitted per unit area normal to the surface of the black body per unit wavelength. With Planck's constant h, vacuum speed of light c, optical wavelength λ, temperature T and Boltzmann constant k the spectral irradiance density is

$$I'(\lambda,T) = \frac{2hc^2}{\lambda^5} \frac{1}{e^{\frac{hc}{\lambda kT}} - 1}. \tag{5.1}$$

Illuminance and irradiance (or alternatively luminous flux and radiant flux) are linked by the luminous efficacy K in $[\text{lm/W}]$. It should be noted that the luminous efficacy depends on the spectral distribution of the light source. It is a measure of how much visible light is generated from a given radiant energy. For a black body at 5900 K the spectral irradiance density in terms of luminous efficacy $K_{\text{black body 5900K}}$ and illuminance H is given in Eq. (5.2). Fig. 5–5 plots $I'(\lambda)$ for 100,000 lx background light illuminance.

$$I'_{BL}(\lambda) = I'(\lambda)\frac{H}{K_{\text{black body 5900K}}}. \tag{5.2}$$

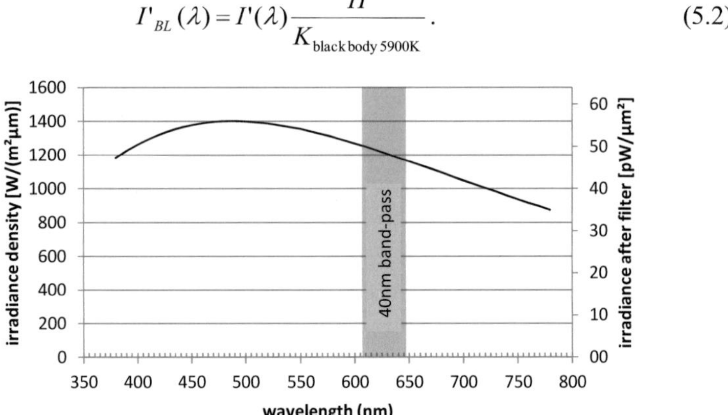

Fig. 5–5: Spectral irradiance density $I'_{BL}(\lambda)$ for a black body at 5900 K at 100,000 lx background light. The right vertical axis denotes the irradiance passing a 40 nm band-pass filter. A filter centered at 635 nm transmits 48 pW/µm².

The optical band-pass filter suppresses most of the background light. Narrow band filters down to 1 nm FWHM are commercially available [100] for high-price laboratory equipment. An inexpensive 40 nm FWHM band-pass for mass-production is used in this work. The background light irradiance I_{BL} after a band-pass centered at the optical carrier wavelength λ_0 with ideal rectangular shape and band-width $\Delta\lambda_{BP}$ is

$$I_{BL}(\lambda_L) = \int_{\lambda_L - \frac{1}{2}\Delta\lambda_{BP}}^{\lambda_L + \frac{1}{2}\Delta\lambda_{BP}} I'_{BL}(\lambda)\,d\lambda. \tag{5.3}$$

5.2.2 Target

Strictly speaking, the target is not part of the optical system of a handheld laser rangefinder. Nevertheless the target is of equal importance as the rest of the optical system as it determines the amount of light seen by the receiving optics. Distances are usually measured normal to the target surface. The amount of light reflected from the target surface depends on the reflectivity R which typically ranges from 10 to 90%. Moreover, the type of reflection is important: diffuse, specular or any combination. On construction sites we are typically dealing with non-cooperative targets with rough surfaces that favor diffuse reflection. The target is therefore modeled as a Lambertian scatterer.

The radiant intensity I emitted from a Lambertian scatterer scales with the cosine of the angle to the surface normal θ with

$$I(\theta) = I_0 \cos(\theta) . \tag{5.4}$$

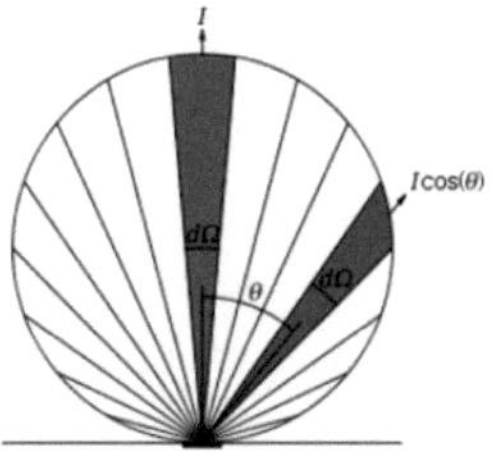

Fig. 5–6: Lambertian scatterer [101]

For a target that is assumed an infinite plain surface, the observed background light irradiance is largely independent of the target distance. While the irradiance from a surface element that radiates into the hemispace in front of the target at distance L decreases with $1/L^2$, the surface area seen by the detector field of view increases with L^2. Both influences cancel.

In various measurement scenarios, background light is incident at an angle γ_{BL} with respect to the surface normal or the target, for example, when light shines through a window at non-normal incidence. The background light irradiance from the target surface then further reduces by $\cos(\gamma_{BL})$.

A retro-reflector can increase the measurement range as it directs the measuring laser light back towards the rangefinder rather than scattering light into the hemispace in front of the target surface. A retro-reflector is beneficial at long distances. However, care must be taken to avoid saturating the detector at short distances (see Section 3.4.3). Moreover, it is time-consuming and not very convenient for a craftsman to mount a reflector at each target. In the following, we assume that no such cumbersome retro-reflector is used at the target.

5.2.3 Transmitter

5.2.3.1 Laser and Modulation

The transmitter comprises the laser driver, laser diode and collimator. High-quality components are available on the market, but the present system aims at achieving a sufficient performance at minimum system cost. Hence, an off-the-shelf edge-emitting laser diode used in mass-production of DVD players and laser pointers is employed as the emitter. In the near future, vertical-cavity surface-emitting lasers (VCSELs) [102] can be expected to further reduce prices. Furthermore, VCSELs with low threshold currents lower the power consumption. A plastic lens collimates the laser beam. Previous laser rangefinders with the same collimator show a beam divergence α of 2 mrad, see Fig. 5–3.

An integrated on-chip laser driver directly modulates the driving current of the laser diode and thus the intensity. A schematic of a transmitter with a directly modulated laser is shown in Fig. 2–3 in Section 2.2.1. The available laser driver that is used in this work does not modulate the laser diode in a perfectly sinusoidal manner with a driving current that is constantly above the laser threshold current but crosses the threshold current. The emitted waveform is also depends on the laser dynamics [103]. The measured spectrum shows a strong modulation at the fundamental modulation frequency (see also Appendix D) but also comprises higher harmonics. The optical output of the laser diode is measured in the experimental section (see Section 6.1.2.1, Fig. 6–4). The measured laser spectrum forms the basis for the input signal to the performance estimation in this chapter (see results in Section 5.4).

Countermeasures against disturbances from higher harmonics of the fundamental modulation frequency include (1) sampling the received signal at the correct sampling frequency with a bandwidth that covers all significant harmonics, (2) filtering the received signal or (3) a combination of filtering and 'oversampling' to cover any significant harmonics in the frequency range that passes the filter. Fortunately, the SPAD detector and subsequent sampling windows provide an inherent filter characteristic as will be shown in more detail in Section 5.3.2.2.

Temperature drift of the laser diode is another important problem for ranging performance. Fig. 5–7 shows that the measured phase drifts over more than a quarter of a modulation period as the temperature changes from -5 to 40 C. The measurement is performed with a current APD-based laser rangefinder at a modulation frequency of 1 GHz. The laser diode temperature is set separately from the circuit board to isolate the warm-up drift of the laser diode. A 90° phase drift at 1 GHz modulation frequency corresponds to 250 ps temporal drift or 37.5 mm distance error.

The optical output at higher temperature lags behind the optical output at lower temperature. Lasing starts after the driving current of the laser driver exceeds the threshold current. As the temperature increases, the lasing threshold current increases. The laser driver reaches a sufficiently strong driving current at a later instant in time and therefore lasing starts later with increasing temperature. A model of the expected temperature-dependent output of the transmitter based on the rate equations of a diode laser can be found in [103].

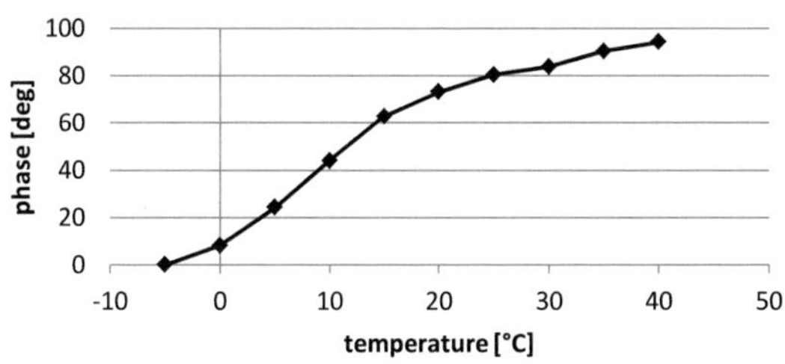

Fig. 5–7: Measured phase over laser diode temperature.

Current APD-based laser rangefinders [4, 5] use a device internal reference path of known length to compensate for any temperature drift of the transmitter. A mechanical flap (see photograph Fig. 2–9 (c)) switches between target and device-internal reference path. Alternatively, a second detector can be used as a reference detector that detects light that has passed a device-internal reference path. The target distance can be determined by comparing the time-of-flight via the target path and the device-internal reference path. The temperature drift of the transmitter is included both in the target path and in the reference path and therefore cancels.

Further alternatively, instead of using a device-internal reference path, the temperature behavior of the transmitter can be estimated [103] to mathematically compensate for any temperature drift of the transmitter.

It should be noted that commercially available devices without internal reference path have a larger specified distance error and a limited temperature range [104].

A second avalanche photodiode (APD) as a reference detector, however, contributes significantly to the cost of a laser rangefinder, as a discrete APD is one of the most expensive components. Nonetheless, high performance laser rangefinders on the market feature a device-internal reference path either with a mechanical flap [4] or a second APD [105].

With CMOS integrated SPADs no additional discrete APD is required since an additional detector can be easily integrated on the same ASIC. The second detector comes at the moderate expense of a slightly increased silicon area of the ASIC. No additional discrete external electronics are required. The absolute temporal drift of a single measurement path reduces to a differential drift, thus reduces to the drift of the temporal difference between target and reference SPAD on the same ASIC.

Finally, also the laser diode wavelength changes with temperature. This drift of, e.g., 0.2 nm/K [106], has to be considered when selecting the minimum spectral filter bandwidth for background light suppression.

5.2.3.2 Burst Modulation

The background light intensity depends on the measurement scenario and can – apart from the optical band-pass filter in the receiver – not be influenced by the design of the laser rangefinder. The light power of the transmitter, however, can be influenced within certain boundaries. Of course, a higher laser power would be highly beneficial to improve the optical signal-to-noise ratio (SNR). For eye-safe operation, the average optical output power of the transmitter is limited to 1 mW by the IEC-60825-1 standard for laser class II [7].

Instead of continuously modulating the laser with 1 mW mean output power, the laser can send out bursts of modulated light. Each burst provides an increased average optical output power of more than 1 mW during a burst on-time τ_{on} which is followed by a burst off-time τ_{off} when the laser is turned off. The average power over the burst period $\tau_{burst} = \tau_{on} + \tau_{off}$ is also limited to 1 mW. The ratio τ_{on} / τ_{burst} defines the burst duty cycle.

Fig. 5–8 sketches the received intensity from a continuously modulated laser (black curve) and a burst modulated laser (grey curve) with same average power on top of background light at unit intensity. Burst modulation improves the optical SNR, because both transmitter and receiver can be switched off during burst off-time τ_{off}. The average signal power remains unchanged at 1 mW, however the average received background light intensity using burst modulation reduces to

$$I_{BL,\text{burst}} = \frac{\tau_{\text{on}}}{\tau_{\text{burst}}} I_{BL} \; . \tag{5.5}$$

Hence, burst modulation increases the optical SNR by reducing the amount of background light seen by the receiver. Burst modulation has not been applied in the performance estimation in Section 5.4, so further accuracy improvements can be leveraged.

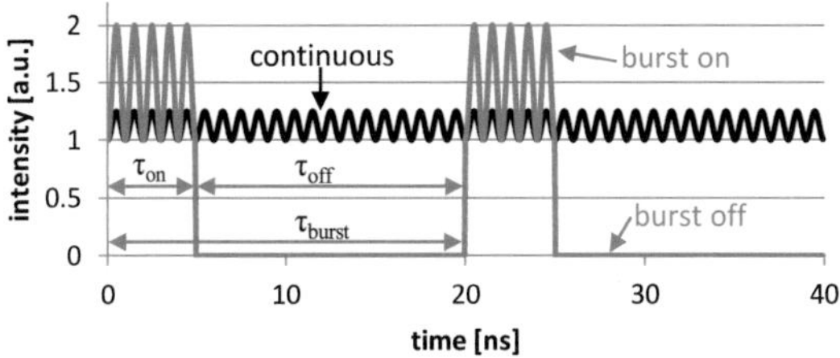

Fig. 5–8: Burst modulation of transmitter. Intensity from background light (BL) is at unity. Black curve: continuously modulated laser and BL. Grey curve: burst modulated laser and BL. The laser is modulated for ¼ of the burst period with a 4 times higher power, hence continuously and burst modulated laser have the same average power. The detector is switched off during 'burst off' times, so that background light is ignored and the overall SNR improved.

It should be noted that the graph in Fig. 5–8 is a schematic to illustrate the principle. It does not cover all limitations set forth by the IEC-60825-1 standard. The standard not only considers the average power but also the energy contained in one burst. Of course the sketched burst on-time of 5 ns is not practical, because the burst on-time must be considerably longer than the round-trip time of the signal light to a distant target and back. A time delay between burst on-time of the transmitter and burst on-time of the receiver can be introduced to capture the entire transmitted burst after the round-trip time to a distant target. More detailed simulations and preliminary measurements for burst modulation in a laser rangefinder can be found in [107]. Considering limitations set forth in IEC-60825-1, a maximum burst period τ_{burst} of 18 µs and a duty cycle >10 % are suggested for practical eye-safe applications.

5.2.3.3 Speckle

Speckles are random intensity patterns resulting from constructive and destructive interference of light reflected from an optically rough surface illuminated by a coherent source. The laser of the rangefinder is a coherent light source that shines onto the target. Coherent light reflected or scattered from the rough surface of the target (rough in the order an optical wavelength) is collected by the receiving optics and causes interference patterns on the detector plane. It should be noted that any change of the direction of the laser, the target surface or imaging system can affect the speckle pattern.

Fig. 5–9 shows an image of a laser spot on a white sheet of paper, hence a rough surface on wavelength scale, imaged with a single-lens reflex (SLR) camera. The intensity variation of the speckle pattern can be clearly seen.

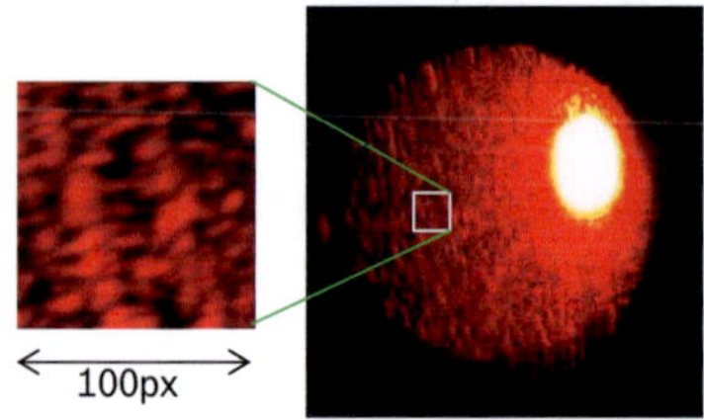

Fig. 5–9: Speckle pattern imaged with a Canon EOS 450D SLR-camera with 5.71 μm × 5.71 μm pixel size and f/#=2.8, f=60mm lens. The speckles are larger than a typical 8 μm diameter SPAD.

It should be noted that speckles only affect the intensity of light from a coherent source, but not from incoherent background light. If the detector active area falls into an interference maximum, the SPAD can see a high count rate which favorably increases the optical SNR. However, too high a count rate can cause SPAD saturation. On the other hand, the detector active area can fall on a speckle minimum, which in turn decreases the count rate and lowers the optical SNR.

There is a difference between conventional APD-based and SPAD-based system. An APD of a current rangefinder has a considerably larger active area of, e.g., 200 μm diameter compared to an 8 μm diameter SPAD. A spatially extended sensor (here: APD) integrates over a larger area with dark and bright speckles, thereby obtaining an average irradiance, whereas a small sensor (here: SPAD) may see a particularly high or low irradiance. Hence a large detector benefits from spatial averaging over multiple speckles. By combining multiple

SPADs, the active area of a SPAD-based system can be increased to artificially create a large detector area which gives an average light level. However, especially at large distances, when signal light is low, the laser spot is concentrated on few SPADs or even a single pixel. It should be noted that a requirement for combining signals from multiple SPADs is a similar timing behavior.

A second option against speckle problems is temporal averaging over multiple different speckle patterns. Recall that any change of the target surface causes a different speckle pattern. In handheld operation, any person trembles ever so slightly when pointing the laser rangefinder at a target a few meters away. Hence the speckle situation changes during measurement time and an average intensity can be measured. Handheld operation favorably mitigates speckle problems.

A third option against speckle problems is emitting short, high peak-power laser pulses with a broad mode spectrum. This effect is mentioned in [108] along with burst modulation.

Speckle reduction with external means for creating different speckle patterns by dithering [109] is possible but not reasonable for a low-cost system.

Please refer to Section 6.1.2.2 for an experimental measurement of the intensity at different speckle situations with a SPAD-based laser rangefinder.

5.2.4 Receiver

A laser range finder for measuring the distance at a single spot on the target can be operated with a low-cost non-imaging optical system. Most notably the system has a fixed focus receiver lens. As shown in Fig. 5–3, the separation s_{parallax} of the optical axis of sending and receiving path defines the parallax. Fig. 5–10 shows a sketch of the system and the respective rays coming from the center of the laser spot on the target at various target distances. In Fig. 5–10 (a) the receiver lens focuses an image of an infinitely distant target on the detector plane. As the target gets closer to the rangefinder in Fig. 5–10 (b) and (c), the laser spot in the detector plane widens and wanders over the detector plane along an axis between the detector plane and a plane spanned by the optical axis of the transmitter and the chief ray of the received light. Hence, the spot position p_{parallax} changes because of the parallax between transmitter and receiver. The spot size, given by the spot diameter d, changes because of defocus.

Three important metrics for a performance estimation of the laser rangefinder are spot size, spot position and irradiance in the detector plane. These metrics are described in the following sections.

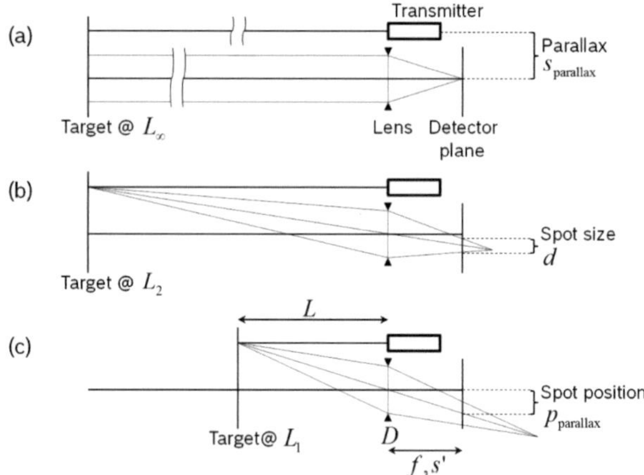

Fig. 5–10: Laser rangefinder with fixed focus receiver lens of diameter D and parallax separation S_{parallax} between transmitter and receiver. In this example the detector plane s' coincides with the focal plane f. (a) Target at infinite distance L_∞: well defined image of laser spot at detector plane. (b) Target at distance L_2: The image of the spot at the detector plane is defocused, hence covers a larger spot size d. The position p of the defocused image of the laser spot moves away from the transmitter (parallax error). (c) Target at distance $L_1 < L_2$: The spot broadens further and the parallax error increases.

5.2.4.1 Spot Size

The minimum spot diameter is given by the diffraction limit of the receiver lens, which can be described with the Airy disk [110]. The Airy disk diameter in terms of optical wavelength λ_0, focal length f and lens diameter D is given by

$$d_{\text{Airy}} = 2.44 \frac{f \lambda_0}{D}.$$

(5.6)

The geometrical spot size in the detector plane depends on the size of the laser spot on the target and on the defocus of the image in the detector plane.

The size of the laser spot on the target, hence the size of the object, translates to a size of the blurred image on the detector plane. We assume a Gaussian beam profile for the transmitter with a beam divergence angle of $\alpha = 2\,\text{mrad}$, defined as the divergence angle where the intensity drops to $1/e^2$. The size of the laser spot on a target at distance L, i.e. the object size, is given by

$$d_{\text{Object}} = \tan(\alpha)L .$$

(5.7)

The magnification M of a lens can be expressed in terms of a ratio of image size d_{Image} and object size d_{Object}. Alternatively, the magnification can be expressed in terms of a ratio of image distance and object distance which are given by the actual position of the detector plane s' and the target distance L,

$$M = \frac{d_{\text{Image}}}{d_{\text{Object}}} = \frac{s'}{L} .$$

(5.8)

The image size represents the diameter of the image of the laser spot in the detector plane. Using Eq. (5.7) and Eq. (5.8) the image size is given by

$$d_{\text{Image}} = d_{\text{Object}} \frac{s'}{L} = \tan(\alpha)L\frac{s'}{L} = \tan(\alpha)s' .$$

(5.9)

The image size is thus independent of the target distance. The reason is that the object size, i.e., the size of the laser spot on the target, increases linearly with the target distance.

The impact of defocus on the spot size is sketched in Fig. 5–10, where the size of the spot in the detector plane increases with shorter target distance from (a) to (c). With the position of the detector plane with respect to the receiver lens s', the focal length f, the receiver lens diameter D and the target distance L, the diameter of the spot because of defocus becomes

$$d_{\text{Defocus}} = D\left(1 - \frac{s'}{f} + \frac{s'}{L}\right) .$$

(5.10)

When the detector plane coincides with the focal plane, the previous expression simplifies to

$$d_{\text{Defocus, s'=f}} = D\frac{f}{L} .$$

(5.11)

With Eq. (5.6), Eq. (5.9) and Eq. (5.10), the combined geometrical and diffraction limited spot size is a convolution of the aforementioned spot diameters given by

$$d_{\text{Spot}} = d_{\text{Airy}} * d_{\text{Image}} * d_{\text{Defocus}}.$$

(5.12)

Fig. 5–11 depicts the spot diameter for the individual contributions as well as the overall spot diameter.

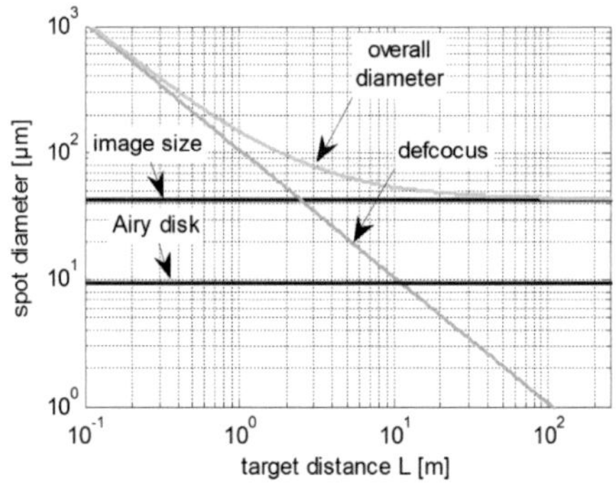

Fig. 5–11: Spot size at detector plane. Airy disk diameter (red), diameter of image of laser spot (blue), diameter due to defocus (green) and the resulting overall diameter (cyan) at the detector plane. Target distance from 10 cm to 250 m.

For distant targets, the amount of light reaching the lens aperture is inherently low. Therefore the system should be designed such that a small concentrated spot is projected onto the array at large target distances. The focus can for example be adjusted for the maximum target distance specified for a laser rangefinder or, as shown in Fig. 5–11, adjusted to infinity.

Nevertheless a small, well-focused image of the laser spot has to be treated with care in a SPAD-based laser rangefinder. The detector is an array of individual SPAD pixels, each comprising a photosensitive active area and non-photosensitive 'dead space' for support circuitry (see also Section 3.2.3.8). Therefore, we introduce 'best-case' and 'worst-case' pixels in the performance estimation. For a worst-case pixel or worst-case detection scenario, sketched in Fig. 5–12 (a), the peak irradiance falls onto the dead space between active areas, so that a SPAD only detects a small amount of light from the periphery of the central laser spot. A best-case pixel or best-case detection scenario, sketched in Fig. 5–12 (b), receives the best-case illumination wherein the maximum amount of signal light of the laser spot falls onto the SPAD active area.

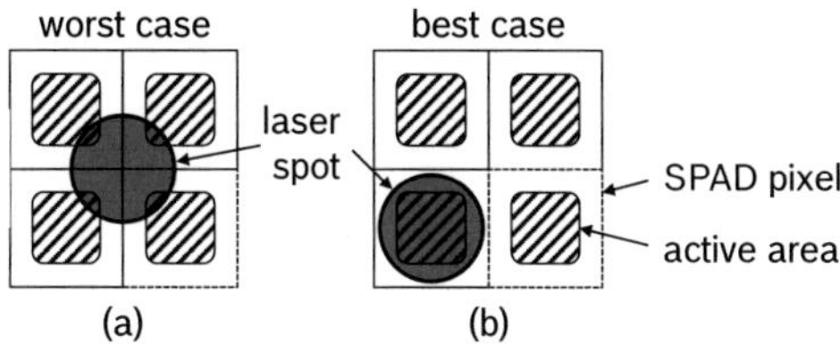

Fig. 5–12: Spot at detector plane with active, i.e., photosensitive, areas. (a) worst-case scenario: the peak irradiance falls in-between active areas, minimum amount of light on the active area; (b) best-case scenario: maximum amount of light an active area.

Lens errors are not considered for now. A certain widening and blurring of the laser spot should be expected from a low-cost plastic receiver lens in practical applications.

5.2.4.2 Spot Position

The position p_{parallax} of the center of the spot in the detector plane with respect to the spot position for a target at infinity (see Fig. 5–10), follows from basic trigonometric considerations. The target distance is denoted by L, the separation of sending and receiving path is denoted by s_{parallax}, and the actual position of the SPAD array (detector plane) with respect to the lens is denoted by s' (see Fig. 5–3). The displacement of the spot on the detector plane, also referred to as the parallax error, is then given by

$$p_{\text{parallax}} = \frac{s_{\text{parallax}}}{L} s'.$$

(5.13)

As can be seen from Eq. (5.13) the parallax error scales with $1/L$. Thus, the parallax error becomes increasingly large at short distances. A large detector area or a large array of SPADs would be needed to cover the entire movement of the spot over the array over distance. A large detector however is expensive.

This problem is solved by introducing a so-called 'near-range element' in the receiver lens. The near-range element is a section of the receiver lens that has a shorter focal length and also redirects some of the light towards the side of the detector array that is oriented towards the transmitter. Thereby, the detector area can be reduced. The actual image of the laser spot wanders off the detector array and only some of the redirected light falls onto the detector array.

Fortunately, the near-range element only has to cover a small part of the receiver lens area. At short target distance, where the parallax error is biggest, the available signal light power is high, so that a fraction of the collected light can yield sufficient signal light for a distance measurement with a good optical SNR. Sophisticated lens geometries with such near-range elements are already employed in state-of-the-art APD-based laser rangefinders [5]. An example of such a lens with near-range element has been disclosed in [111] (Bosch patent application).

Fig. 5–13 shows the results of a ray tracing simulation [112] using such a lens. The white circles in Fig. 5–13 represent the active areas of individual SPADs. For a target at 20 m distance, Fig. 5–13 (a), the spot is well focused on the detector plane, hence concentrates the signal light on a small area. At 1 m target distance, Fig. 5–13 (b), the spot broadens because of defocus. The transmitter is at the right side of the detector plane with a certain parallax. Because of this parallax the spot wanders over the detector plane away from the transmitter as described in Eq. (5.13) and indicated by the white arrow. Fig. 5–13 (b) also shows the effect of the near-range element. A portion of the signal light is redirected to the side of the detector plane facing the transmitter. At a few centimeters target distance (not shown in Fig. 5–13), the center of the laser spot does not fall on the detector array any more, but wanders off to the side opposite to the transmitter. In that case only light from the near-range element is detected. Fortunately, the respective signal light irradiance at short target distances yields sufficient intensity for distance measurement.

For a distance measurement simulation, always the SPAD with the highest irradiance is selected. In case of saturation (see Section 3.4.3) at short distances a different SPAD in the periphery of the spot with lower intensity can of course be evaluated.

A SPAD array provides further benefits in addition to enabling the selection the SPAD with highest irradiance. For a low-cost system not only the cost of the individual components but also the cost of the manufacturing process is crucial. At the price of a larger silicon area, a larger SPAD array can be integrated into the ASIC. With a large SPAD array the alignment tolerances in the manufacturing process can be met – which can save expenses for lens alignment robots and allow pick-and-place automation in the manufacturing of the low-cost laser rangefinder.

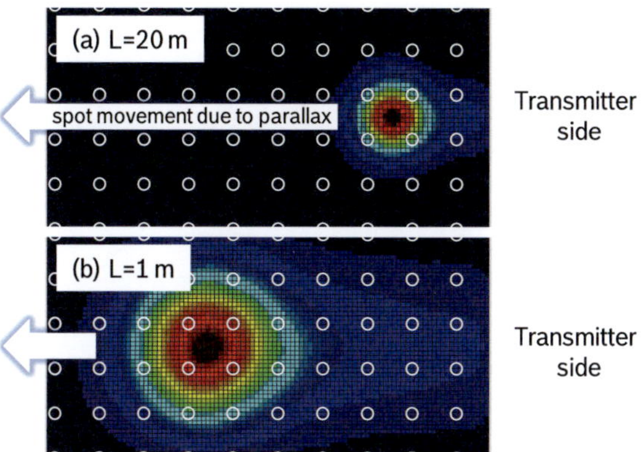

Fig. 5–13: Laser spot on SPAD array in detector plane. White circles indicate outlines of SPAD active areas. (a) 20 m target distance: well concentrated spot. (b) 1 m target distance: broadened spot because of defocus, different color scale with 60× higher peak irradiance. At shorter target distance the spot moves away from the transmitter side because of parallax. The near-range element of the receiver lens redirects a portion of the received signal light towards the transmitter. Both simulation show a 'worst-case' scenario, where the peak of the laser spot falls in-between four SPADs and reduces the maximum detectable irradiance on the SPAD active area.

5.2.4.3 Irradiance at Detector Plane

The spot size on the detector plane determines the area over which the light received from the target at a certain distance is distributed. Fig. 5–14 shows a graph of the irradiance on the detector plane as a function of the target distance. At short distances, the image of the laser spot is defocused, hence the irradiance distributes over a larger detector area, and the curve is rather flat. At far distances, the image of the laser spot is focused onto the detector plane, so the irradiance drops with the expected $1/L^2$. A larger lens diameter directly increases the irradiance on the detector plane.

The calculation in Fig. 5–14 is based on the geometric spot size model. For more complex lens geometries, in particular lens geometries including a near-range element [111], a ray-tracing simulation such as Zemax can provide more accurate results.

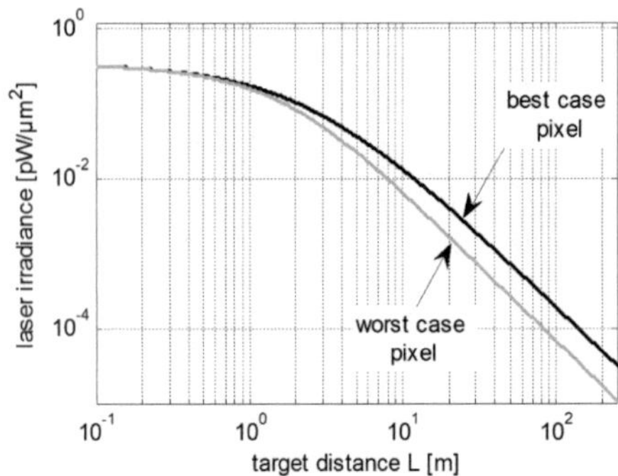

Fig. 5–14: Irradiance of image of laser spot at detector. Irradiance on a SPAD at the center of the laser spot (best-case pixel) and for a SPAD in the periphery, when the laser spot falls in-between SPAD active areas (worst-case pixel).

5.2.5 Optical Crosstalk

Two separate photodetectors for target and reference measurement are suggested to compensate for the temperature drift of the laser diode (see Section 5.2.3.1). In conventional APD-based laser rangefinders these two detectors are separate hardware elements, one in the sending and one in the receiving optical path [105]. With two sufficiently spaced photodetectors, optical crosstalk is negligible. However, if the photodetectors are integrated in an ASIC, the distance shrinks to some hundred micrometers and optical crosstalk becomes an issue.

Separate ASICs for target and reference path are also possible in a SPAD-based system, however, for cost reasons a single ASIC is preferred. Fig. 5–15 shows an embodiment of target and reference detector on a single ASIC with crosstalk from internal reflections and scattering. The optical separation between target and reference path is crucial. We present a patent-pending solution in [19]. The performance estimation in this chapter considers optical crosstalk as an additional error phasor of fixed magnitude. Optical crosstalk occurs both from target to reference as well as from reference to target.

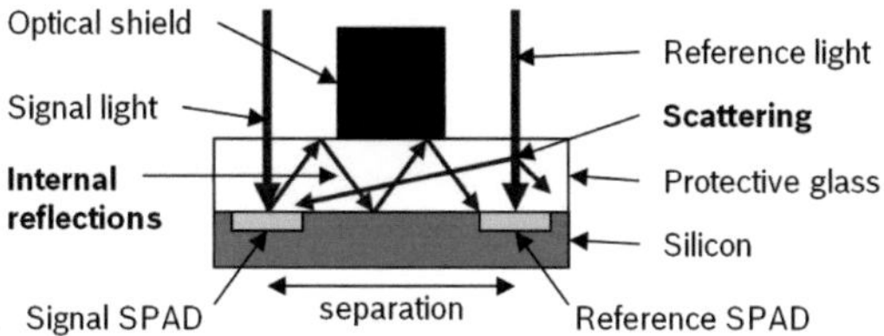

Fig. 5–15: Laser rangefinder with target and reference detector on a single ASIC.

5.3 Electronics

The core electronic elements for a laser rangefinder are laser diode with driver, detector and timing circuitry. The laser has been briefly described as a part of the transmitter in the previous section. Laser diode and driver will not be analyzed in detail as they have not been modified over state-of-the-art ranging systems.

Chapters 3 and 4 describe and characterize single-photon avalanche diodes (SPADs) in detail. Therefore the present section only briefly refers to parameters that have to be considered for distance measurements using the single-photon synchronous detection (SPSD) concept. The focus is set on the timing circuitry. The timing circuitry comprises a so-called 'binning architecture' or simply 'binning' which samples incident pulses from the SPAD in synchronism with the modulation of the laser. The sampling windows are referred to as 'bins'.

In 3D ranging cameras each pixel has a memory and a distance evaluation unit in order to measure the distance for each pixel. In contrast to 3D cameras, the proposed single-spot laser rangefinder only measures the distance at a single spot on the target, i.e., the distance between rangefinder and target. Hence, only one distance evaluation unit is required in the most basic configuration. An optional additional distance evaluation unit enables a parallel reference measurement via a device-internal reference path to compensate for drifts of the transmitter (see Section 5.2.3.1). The distance evaluation circuit can be placed outside the array and can be shared among all pixels. This reduces the required silicon area and improves the fill factor (see Section 3.2.3.8) of the SPAD array. Moreover, circuitry placed outside the array allows more degrees of freedom in terms of the number of bins and the size of counters to count individual SPAD pulses.

Error contributions, such as photon noise, binning non-idealities, crosstalk and delay times, which impair any distance measurement with the proposed system, will be described herein. The error contributions are considered further below in the performance estimation in Section 5.4.

The target of the present work is a low-cost laser rangefinder. Hence, any error correction, which would require costly hardware for extensive signal processing such as a digital signal processor (DSP), is ruled out up front. Instead, we propose a bin homogenization scheme in Section 5.3.5, which enables bin width correction directly within the ASIC. No further external signal processing for bin width correction is required with the proposed bin homogenization scheme.

5.3.1 SPAD

So far, background light and signal light have been discussed in terms of intensity or irradiance, i.e., in terms of a power per area (see Fig. 5–14). The SPAD detector converts incident photons into single electrical pulses. The detected pulse rate is proportional to the intensity incident on the detector active area A_{det}. We denote the time-varying signal intensity from the modulated signal light which is backscattered from the target with $I_S(t)$, the background light intensity with I_{BL}, Planck's constant with h, the speed of light with c, the optical frequency with $f_0 = c / \lambda_0$, the photon detection efficiency with η_{PDE} (see Section 3.2.3.1) and the dark count rate with n_{DCR}. For these parameters, the ideal count rate $m(t)$ of detected photons and dark counts is

$$m(t) = \left[I_S(t) + I_{BL} \right] A_{\mathrm{det}} \frac{1}{h f_0} \eta_{\mathrm{PDE}} + n_{DCR} . \tag{5.14}$$

The actual count rate n follows from the ideal count rate m depending on the saturation behavior of the SPAD and its quenching circuit (see Section 3.4.3). The simple formula 'the more SPAD events the better' is not true, as saturation effects can reduce the actual count rate n below the ideal count rate m.

Fig. 5–16 shows calculated count rates of a SPAD based on the irradiances for best-case and worst-case pixel in Fig. 5–14 on page 146. Dark counts and background light create an offset in addition to counts from signal light. A maximum count rate of 11.5 MHz for signals coming from small target distances decreases to 1 MHz at long distances, in which case the count rate is dominated by counts from background light.

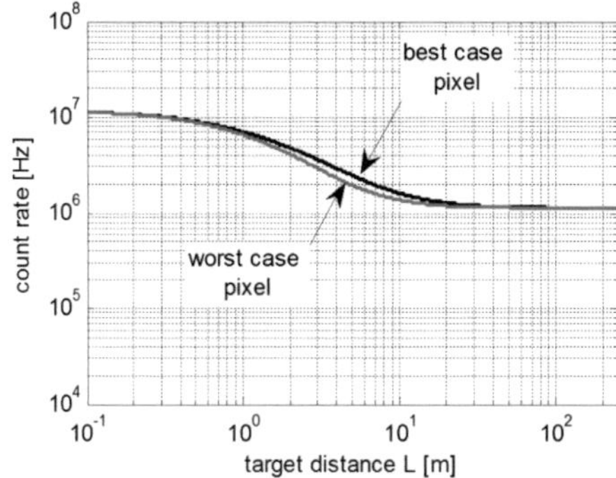

Fig. 5–16: Count rate of a SPAD pixel as a function of distance. A best-case pixel is a SPAD at the center of the laser spot. For a worst-case pixel, the laser spot falls in-between the active area of four SPADs in the array.

Average count rates are presented in Fig. 5–16. The actual number of photons, which in turn cause an electrical output pulse, depends on the emission characteristic of the light source as described in Section 3.2.2 on photon statistics.

The influence of SPAD jitter on the received signal is discussed together with the influence of the bin width, i.e., the sampling widow width, in Section 5.3.2.2.

5.3.2 Binning

The binning concept for sampling the received signal has been introduced in Section 2.3.3 (page 31). The concept of operating the binning in synchronism with the laser modulator for frequency-domain reflectometry with SPADs has been introduced in Section 2.3.3.2 (page 35).

5.3.2.1 Ideal Binning

An ideal binning for a SPAD-based laser rangefinder comprises sampling windows (bins) that are equidistantly distributed over the modulation period. Moreover, the bins are adjacent to each other and cover the entire period, so that no SPAD event is lost. Appendix C provides further information about sampling a received signal.

Fig. 5–17 (a) shows the light intensity incident on the SPAD active area as a continuous sinusoidal curve for one modulation period (black curve). The SPAD provides a count rate that is proportional to the intensity. The SPAD receives both signal light and background light (including SPAD dark counts). In the shown example, the signal light is a sine wave with a modulation depth of $m_{\mathrm{signal}} = 100\%$. Background light and SPAD dark counts form a constant offset level.

Fig. 5–17 (b) shows an ideal binning for sampling the received signal. In this example, the binning comprises eight bins (bin 0 to bin 7). Each of the bins is defined by its opening and closing edges $t_{\mathrm{open,k}}$ and $t_{\mathrm{close,k}}$. The time between opening and closing edges defines the bin width b. SPAD events $s_{\mathrm{Rx}}(t)$ that fall in-between the opening and closing edges during each modulation period l of duration T increment the counter values x_k,

$$x_k = \int_{t_{\mathrm{open,k}}+lT}^{t_{\mathrm{close,k}}+lT} s_{\mathrm{Rx}}(t)dt \quad , \quad l \in N_0 .\tag{5.15}$$

SPAD counts falling into bin 0 increase counter 0, SPAD counts falling into bin 1 increase counter 1, and so on. The binning samples the incident signal by periodically repeating the sequence of bins 0 to 7 in synchronism with the laser modulation. A measurement over multiple periods builds up a histogram of counts per bin as shown in Fig. 5–17 (c). The respective counter values represent the received signal.

With an ideal binning as shown in Fig. 5–17 (b), where all bins have the same width, the constant level from background light and dark counts is homogeneously distributed over all bins (dark grey). The histogram portion representing modulated signal light is depicted on top (light grey).

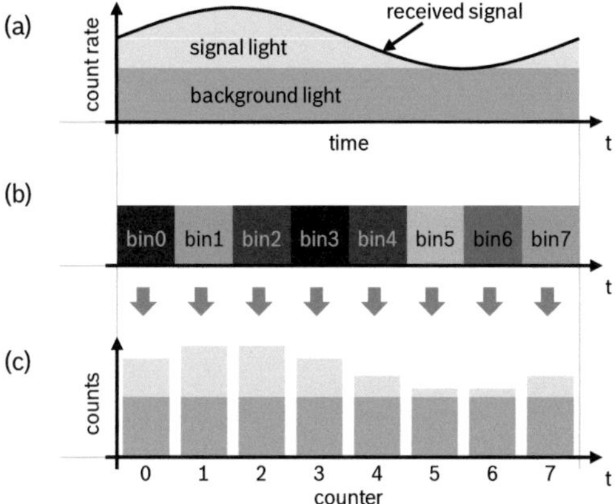

Fig. 5–17: Ideal binning. (a) Sinusoidal signal with offset from background light and SPAD dark counts. (b) Ideal bins are equidistantly distributed over the modulation period. No gaps or overlaps between neighboring bins. (c) The received signal is represented by a histogram of SPAD counts in each bin.

As described in introductory chapter, Section 2.2, the target distance is calculated from the phase difference φ between sent and received signal from the receiver phase Φ_{Rx} and transmitter phase Φ_{Tx}. As an alternative to the transmitter phase, the measured phase of a portion of the transmitted signal light that has passed a reference path of known length can be used. The receiver phase Φ_{Rx} can be computed from the counter values for bins 0 to 7 by means of a discrete Fourier transform (DFT), Eq. (5.16) as described in Section 2.3.2. The counter values are denoted $x_{0...7}$. The Fourier coefficients are given by

$$X_l = \frac{1}{N}\sum_{k=0}^{N-1} x_k e^{-j2\pi\frac{kl}{N}} \quad , \quad N = 8 \ , \ l = 0,1,...,N-1 . \tag{5.16}$$

The DC offset is given by the 1st element of the Fourier transform X_0, the phase and amplitude of the fundamental modulation frequency by the 2nd complex element X_1. The phase of the fundamental modulation frequency used for calculating the measured distance is given by the argument of X_1,

$$\Phi_{Rx} = \arg(X_1). \tag{5.17}$$

The modulation depth or modulation index at the modulation frequency of the detected signal after sampling is represented by the coefficient

$$m_M = \frac{|X_1|}{|X_0|}.$$ (5.18)

Examples of modulation depths that can be achieved with different modulation schemes are shown in Appendix D.

Fig. 5–18 shows another histogram of counter values and further illustrates the reconstructed received signal that has been reconstructed from the measured counter values. The phase of the received signal is denoted as the receiver phase Φ_{Rx}.

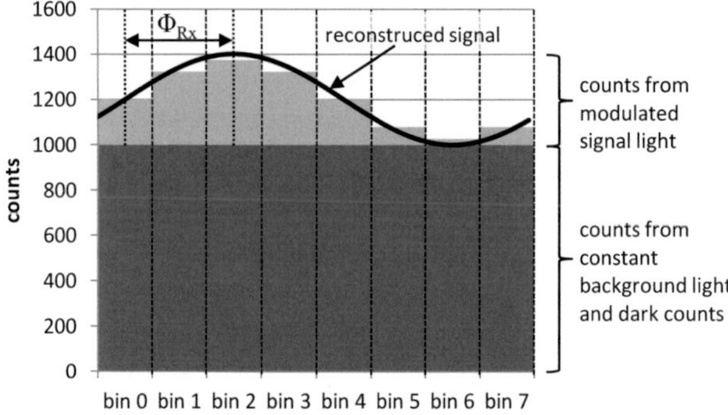

Fig. 5–18: Histogram of counts detected with ideal bins of same widths. The received signal can be reconstructed from the histogram. The target distance can be determined based on the measured receiver phase Φ_{Rx} and the known transmitter phase Φ_{Tx} (not shown).

An ideal binning procedure for a laser rangefinder samples the incident signal with an integer multiple of bins N per period. For a measurement at different frequencies, i.e., to resolve distance ambiguities (see Section 2.2.3), the bin width is adjusted to cover $1/N$ of the period regardless of the modulation frequency. This approach simplifies signal processing as the evaluated modulation frequency always corresponds to a frequency of the discrete Fourier transform. Hence there is no need for windowing which allows easy signal processing.

Since the number of samples is advantageously a power of two, one can conveniently use existing FFT algorithms. The limited sample number with only eight counter values allows implementation in a low-cost microcontroller with limited computational power. Phase calculation using correlation filters is an alternative implementation that does not require the number of bins to be a power of two. However, DFT and correlation filters are mathematically equivalent.

5.3.2.2 Filter Characteristics of Bin Width and SPAD Jitter

The laser rangefinder presented in this work measures distances in the frequency domain. The filter characteristics of the binning or of SPAD parameters could influence a distance measurement.

A bin is a rectangular sampling window of width b that acts as a filter on the received signal. The following definitions are used:

$$\text{rect}\left(\frac{t}{b}\right) = \begin{cases} 1/b, & |t| < b/2 \\ 0, & |t| > b/2 \end{cases}$$
$$\text{sinc}(bf) = \frac{\sin(\pi b f)}{\pi b f}.$$

$$(5.19)$$

The filter characteristic of the sampling process is described by the impulse response $h_{\text{bin}}(t)$ in time domain and the transfer function $H_{\text{bin}}(f)$ in frequency domain [91],

$$h_{\text{bin}}(t) = \text{rect}(b) \;\Rightarrow\; H_{\text{bin}}(f) = \text{sinc}(bf).$$

$$(5.20)$$

Thus, under the condition that the bin width is smaller than the modulation period $T = 1/f_m$, a variation of the bin width does not cause a phase error.

This sinc filter characteristic in frequency domain can also be shown starting from Eq. (5.15) using a sinusoidal signal $s_{Rx}(T) = \sin(2\pi f_m t)$ with modulation frequency f_m and bin width b,

$$\frac{1}{b}\int_{\tau-b/2}^{\tau+b/2} \sin\left(2\pi\frac{t}{T}\right)dt = \left[-\frac{1}{b}\frac{T}{2\pi}\cos\left(2\pi\frac{t}{T}\right)\right]_{\tau-b/2}^{\tau+b/2}$$

$$= \left(\frac{1}{b}\frac{T}{2\pi}\right)\left[-\left\{\cos\left(2\pi\frac{\tau}{T}\right)\cos\left(2\pi\frac{b/2}{T}\right)-\sin\left(2\pi\frac{\tau}{T}\right)\sin\left(2\pi\frac{b/2}{T}\right)\right\}\right.$$

$$\left.+\left\{\cos\left(2\pi\frac{\tau}{T}\right)\cos\left(2\pi\frac{-b/2}{T}\right)-\sin\left(2\pi\frac{\tau}{T}\right)\sin\left(2\pi\frac{-b/2}{T}\right)\right\}\right] \qquad (5.21)$$

$$= \frac{1}{2}\frac{T}{\pi b}2\sin\left(2\pi\frac{\tau}{T}\right)\sin\left(\frac{\pi b}{T}\right)$$

$$= \mathrm{sinc}\left(\frac{b}{T}\right)\sin\left(2\pi\frac{\tau}{T}\right).$$

Furthermore, the SPAD jitter acts as a filter on the received signal. SPAD jitter has been defined as the statistical time variation between true photon arrival and actual SPAD output pulse (see Section 3.2.3.6). The average measured arrival time from a plurality of SPAD output pulses is a weighted average of the measured arrival times of individual SPAD output pulses. The weighting function is the probability density distribution of the jitter [113, 114]. This weighted averaging can also be described as a convolution of a time-domain signal with the probability density function (PDF) of the jitter [113, 114]. Assuming a Gaussian PDF with standard deviation $\sigma_{\Delta t,\mathrm{jitter}}$, we obtain the impulse response $h(t)_{\mathrm{jitter}}$ and the transfer function $H(f)_{\mathrm{jitter}}$ of the jitter low-pass filter [114],

$$h_{\mathrm{jitter}}(t) = \frac{1}{\sigma_{\Delta t}\sqrt{2\pi}}e^{-t^2/2\sigma_{\Delta t}^2} \quad \Rightarrow \quad H_{\mathrm{jitter}}(f) = e^{-2(\pi\sigma_{\Delta t}f)^2}. \qquad (5.22)$$

The combined influence of SPAD jitter and binning is a convolution in time domain corresponding to a multiplication in frequency domain,

$$H_{\mathrm{combined}}(f) = H_{\mathrm{bin}}(f)H_{\mathrm{jitter}}(f). \qquad (5.23)$$

Fig. 5–19 shows the transfer functions for a modulation frequency of 1 GHz, eight sampling windows (bins) per period, each bin having a bin width of 125 ps, and 180 ps FWHM ($\sigma_{\Delta t,\mathrm{jitter}} = 180\,\mathrm{ps}/2/\sqrt{2\ln 2} = 76.4\,\mathrm{ps}$) Gaussian SPAD jitter. The transfer function of the sampling depends on the bin width which in turn depends on the modulation period and the number of bins per period. SPAD jitter is a device parameter that is independent of the binning. At a frequency of 1 GHz, the value of the combined transfer function is 86.8 %.

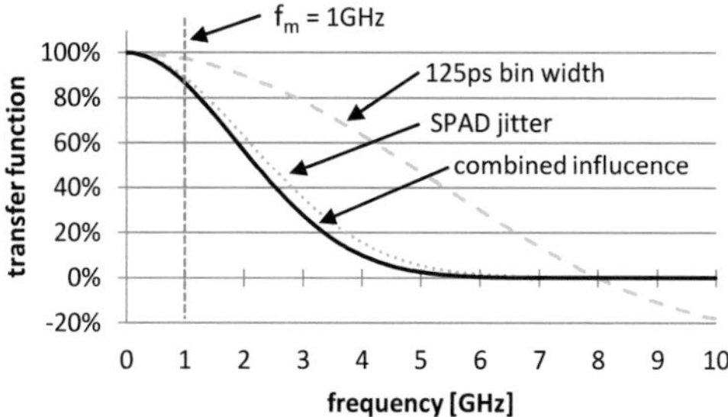

Fig. 5–19: Filter characteristic due to SPAD jitter and bin width. 180ps FWHM Gaussian SPAD jitter. A modulation frequency of 1 GHz and sampling with 8 ideal bins per period gives a sampling window width of 125 ps.

Measurements of the optical output signal of the laser have shown a strong presence of higher harmonics in the optical signal (see Fig. 6–4). As a consequence, a certain amount of SPAD jitter is actually desirable as its low-pass filter characteristic helps to reduce the disturbance from higher harmonics. Aliasing of higher harmonics influences the signal at the fundamental modulation frequency as illustrated in Fig. 5–20. With 8 bins per period, i.e., with a sampling rate of 8 times the fundamental modulation frequency of 1 GHz, the 7th (7 GHz) and 9th (9 GHz) harmonic are the first higher harmonics that contribute to the fundamental modulation frequency.

Fortunately, the filter characteristic due to SPAD jitter and bin width shown in Fig. 5–19 attenuates the 7th and 9th harmonics of the fundamental modulation frequency (and also any further higher harmonics) by more than three orders of magnitude.

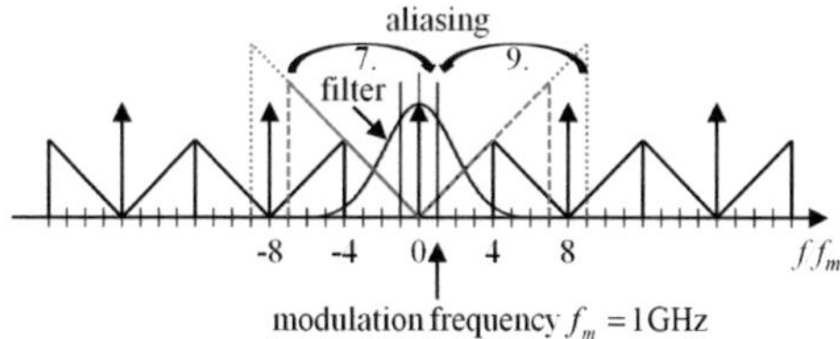

Fig. 5–20: Spectrum with aliasing. With 8 bins per period, i.e., sampling the signal at 8 times the fundamental modulation frequency of 1 GHz, the 7th (7 GHz) and 9th (9 GHz) harmonic are the first that can contribute to the fundamental modulation frequency. However, these harmonics are already attenuated by more than three orders of magnitude by the filter characteristic due to SPAD jitter and bin width (see **Fig. 5–19**).

For comparison, Fig. 5–21 shows the transfer functions for a modulation frequency of 1 GHz, four sampling windows (bins) per period, each bin having a bin width of 250 ps, and 180 ps FWHM Gaussian SPAD jitter. At a frequency of 1 GHz, the value of the combined transfer function is 80.2%. Assuming 4 bins per period, i.e., a sampling rate of 4 times the fundamental modulation frequency of 1 GHz, the 3rd (3 GHz) and 5th (5 GHz) harmonic are the first higher harmonics that contribute to the fundamental modulation frequency. At a frequency of 3 GHz, the value of the combined transfer function is still at 10.6 % and therefore does not sufficiently suppress disturbances due to higher harmonics.

Hence, the binning with eight bins per period is selected for the SPAD-based laser rangefinder.

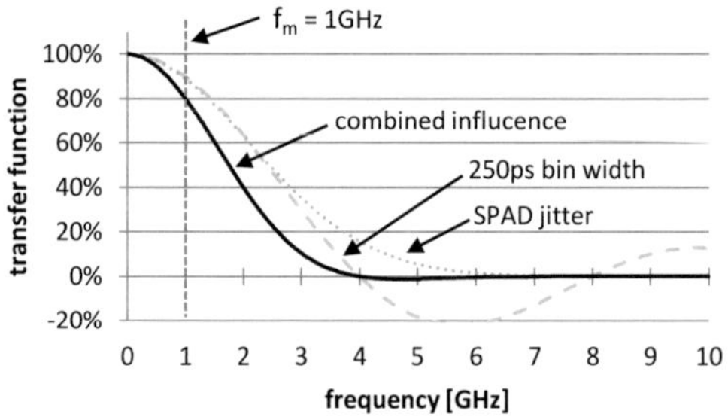

Fig. 5–21: Filter characteristic due to SPAD jitter and bin width. 180ps FWHM Gaussian SPAD jitter. A modulation frequency of 1 GHz and sampling with 4 ideal bins per period gives a sampling window width of 250 ps.

5.3.2.3 Modulation of Received Signal

The detected signal comprises SPAD counts due to modulated signal light, background light and dark counts (see Fig. 5–18 on page 152). The modulation index on the fundamental frequency f_m of the detected signal after sampling is denoted with m_M (see Eq. (5.18)). The modulation depth m_M not only depends on the background light, dark counts, signal light and the modulation depth of the signal light but also on the filter characteristic due to SPAD jitter and bin width (see previous Section) which further attenuates the modulation.

We denote the average dark count rate with n_{DCR}, the average background light count rate with n_{BL}, the average signal light count rate with n_{signal}, the modulation depth of the signal light with m_{signal} and the measurement time with

t_{meas}. The value of transfer function at the fundamental modulation frequency of the signal f_m is denoted with $H_{\text{combined,fm}}$ (see Eq. (5.23)). For these parameters, the modulation depth is

$$m_M = \frac{H_{\text{combined,fm}}\, m_{\text{signal}}\, n_{\text{signal}}\, t_{\text{meas}}}{(n_{\text{signal}} + n_{BL} + n_{\text{DCR}})\, t_{\text{meas}}}. \tag{5.24}$$

The term $(n_{\text{signal}} + n_{BL} + n_{\text{DCR}})\, t_{\text{meas}}$ in Eq. (5.24) gives the average number of detected photons $\overline{N}$ during the measurement time t_{meas} and thereby also gives an indication of how much photon noise can be expected. As described in Section 3.2.2, the photon noise for a Poisson distributed light source is given by the square root of the average number of photons, $\sigma_N = \sqrt{\overline{N}}$. We can therefore define the signal-to-noise ratio (SNR) for the case dominated by background light with

$$SNR^{(p)} = \frac{H_{\text{combined,fm}}\, m_{\text{signal}}\, n_{\text{signal}}\, t_{\text{meas}}}{\sqrt{(n_{\text{signal}} + n_{BL} + n_{\text{DCR}})\, t_{\text{meas}}}} = \frac{m_M\, \overline{N}}{\sigma_N} = m_M \sqrt{\overline{N}}. \tag{5.25}$$

The $SNR^{(p)}$, as defined here, gives a ratio of photons of the modulated signal light to photon noise. The $SNR^{(p)}$ scales with $\sqrt{\overline{N}}$ and thus with $\sqrt{t_{\text{meas}}}$.

An alternative definition of the SNR is commonly used when a photocurrent is measured. A photodetector generates a photocurrent i_{photo} that is proportional to the received light power $P_{\text{opt}} \sim i_{\text{photo}}$. The photocurrent is measured via a measuring resistance R_{meas}. The electrical power at the measuring resistance is $P_{el} = R_{\text{meas}} i_{\text{photo}}^2$, and therefore $P_{el} \sim P_{\text{opt}}^2$. For this case, the SNR scales with $\overline{N}$,

$$SNR^{(el)} = \frac{m_M^2\, \overline{N}^2}{\sigma_N^2} = m_M^2\, \overline{N}. \tag{5.26}$$

However, since the SPADs provide individual electrical output pulses for detected photons, and because these individual pulses are processed by the subsequent circuitry, the definition of Eq. (5.25) is used herein.

Fig. 5–22 shows the modulation depth m_M versus the target distance for the count rates previously shown in Fig. 5–16. The difference in modulation depth between best-casc and worst-case pixel (see Fig. 5–12) illustrates the problem of the laser spot falling in-between different active areas of a SPAD array. At a target distance of 20 m and background light of 10,000 lx, the modulation depth has dropped down to 4.6 % for the worst-case pixel compared to 10.2 % for the best-case pixel. Hence, at such target distances the signal to be sampled with the binning architecture is strongly dominated by the non-modulated background light.

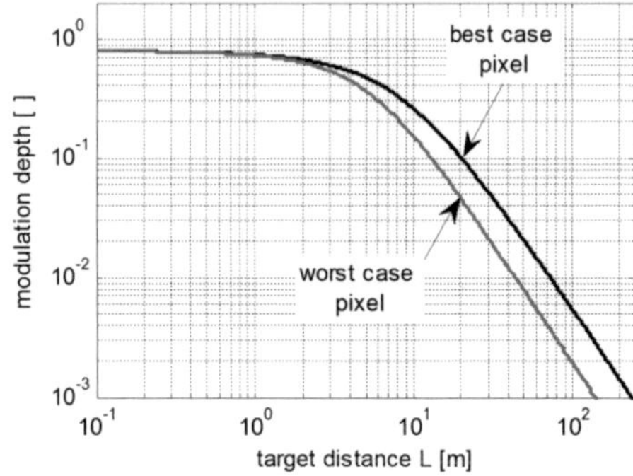

Fig. 5–22: Modulation depth vs. target distance. At 20 m target distance and 10,000 lx background light, the modulation depth on the fundamental frequency is 10.2 % for the best-case pixel, and down to 4.6 % for the worst-case pixel.

5.3.2.4 Distance Error due to Photon Noise

The number of photons falling into each bin is a Poisson distributed random variable, as described in Section 3.2.2. Hence, the number of SPAD events falling into each bin has a standard deviation equal to the square root of the average number of SPAD events falling into the respective bin. For each bin, the number of SPAD events is represented by a corresponding counter value. Fig. 5–23 illustrates the counter values for eight bins with error bars that indicate the standard deviation. Of course Fig. 5–23 exaggerates the impact of photon noise, as unrealistically low counter values are shown for illustration. With the calculated count rates (see Fig. 5–16) in the MHz region and measurement times of up to 4 seconds the relative photon noise (see Eq. (3.13)) is significantly lower.

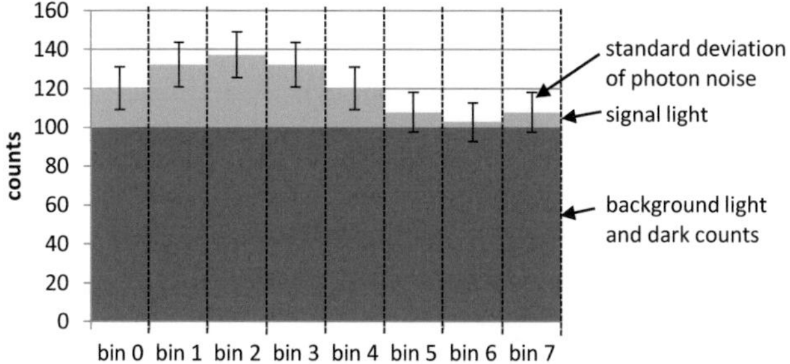

Fig. 5–23: Average counter values with photon noise. Error bars indicate the standard deviation for Poisson distributed counts.

Photon noise is present on any detected counter value. Not average but these noisy counter values in turn are the basis for calculating the phase shift between transmitter and receiver – from which the measured distance is calculated. The photon noise, in particular due to background light, provides the ultimate limit [49] for distance measurement accuracy. Using Eq. (5.25), the phase error from photon noise for the background light dominated case is given by

$$\Delta\Phi_{\text{photon noise}} = \frac{\sqrt{2}}{SNR^{(p)}} .$$

(5.27)

As can be seen from Eq. (5.27), the phase error from photon noise is independent of the modulation frequency, but reduces with a better signal-to-noise ratio. The distance error for a given phase error $\Delta\Phi_{\text{photon noise}}$ due to photon noise is calculated using Eq. (2.10),

$$\Delta L_{\text{photon noise}} = \frac{1}{2} c \frac{\Delta\Phi_{\text{photon noise}}}{\omega_m} .$$

(5.28)

Hence the distance error for a fixed phase error scales with $1/\omega_m$ and is thus inversely proportional to the modulation frequency $f_m = \omega_m / 2\pi$. We can exploit this property by choosing the modulation frequency f_m as high as possible to reduce the distance error.

The distance error from photon noise as a function of the modulation frequency is depicted in the results section of the system performance estimation in Section 5.4, page 194, Fig. 5–46.

5.3.3 Binning Implementations and Non-Idealities

A stable, high modulation frequency is desired for distance measurement by frequency-domain reflectometry. Within an ASIC, such a stable, high measurement frequency for driving both the laser modulator as well as the binning architecture can be generated using a voltage controlled oscillator (VCO) that is embedded in a phase locked loop (PLL). The PLL is locked to a stable external reference frequency, for example a 10 MHz quartz oscillator that can be shared among microcontroller and ASIC.

Fig. 5–24 shows a block diagram of a phase-locked loop (PLL) [115]. The PLL comprises a pulse-fraction discriminator (PFD), a charge pump (CP), a loop filter (LF), a voltage-controlled oscillator (VCO), a clock buffer and a divider in the feedback path. The reference frequency RCLK is provided from an external stable oscillator. The pulse-fraction discriminator compares RCLK with the signal from the feedback path and controls the charge pump. A loop filter (LF) filters the output signal of the CP to provide the voltage-controlled oscillator (VCO) with a control voltage (VCNT) that sets the frequency of the VCO. A clock buffer buffers the RF output signal (CLK) of the VCO and feeds the output signal back through a divider 1/N for comparison in the PFD.

To generate an output frequency of 1 GHz from a reference frequency of 10 MHz, the divider N is set to N=100. Hence, the PFD compares the 10 MHz reference signal with the 1 GHz frequency divided down to 10 MHz. In the implementation shown in Fig. 5–24, the PFD comprises two flip-flops and an AND-gate. If the signal from the reference path lags behind RCLK, the output of the upper flip-flop is at high level, which opens the upper switch of the charge pump and thus increases the voltage VCNT. When the signal from the reference path triggers the lower flip-flop of the PFD, its output switches to high level. The AND-gate resets both flip-flops in the PFD and opens the switches of the charge pump. In response to the higher control voltage, the frequency of the VCO increases. This reduces the delay between RCLK and the signal from the feedback path. The VCO is finally locked to the reference frequency. A different frequency can easily be set by choosing a different divider value.

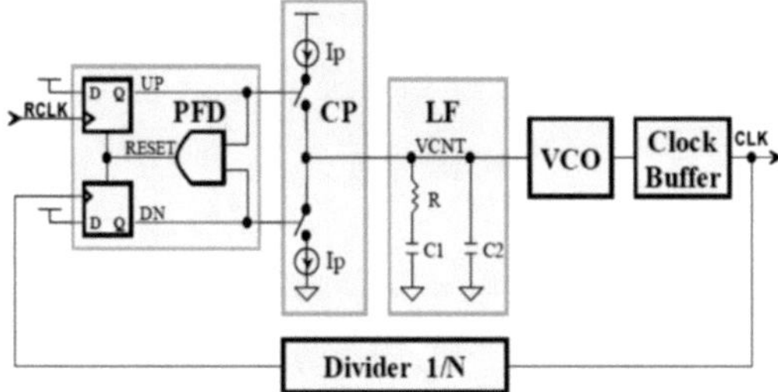

Fig. 5–24: Phase-locked loop (PLL). The reference frequency RCLK is provided by an external stable oscillator. The pulse-fraction discriminator (PDF) compares the reference frequency with the signal from the feedback path and controls the charge pump (CP). A loop filter (LF) filters the output signal of the CP to provide the voltage-controlled oscillator (VCO) with a control voltage (VCNT) that sets the frequency of the VCO. A clock buffer buffers the RF output signal (CLK) of the VCO and feeds the output signal back through a divider 1/N for comparison in the PFD. © 2006 [115]

Two typical options for VCO design are an LC-oscillator and a ring oscillator [116]. For high-quality RF generation, an LC-oscillator is preferred because of its lower phase noise [115, 117]. A ring oscillator, however, is less expensive. The impact of phase noise on the distance measurement error can be seen analogous to SPAD jitter (Section 3.5.1).

5.3.3.1 Sorting SPAD Pulses into Counters

The binning receives electrical pulses from a SPAD or SPAD array and ensures that these pulses increment the correct counter. The general concept has already been explained in Section 2.3.3.2 with reference to Fig. 2–16.

An implementation for sorting SPAD pulses into counters is shown in Fig. 5–25. The input of each of the counters 0 to 7 is connected to the output of a SPAD array. Each of connection line that connects a counter to the SPAD output comprises a switch. The switches are controlled by control signals bin<0> to bin<7>. When the control signal is at high logic level, the switch closes and establishes a connection between the input of the counter and the output of the SPAD array. Thereby, the control signals bin<0> to bin<7> determine the time windows during which SPAD pulses can increment counters 0 to 7. The control signals thus determine the sampling windows for sampling the received signal.

The control signals are illustrated next to the control lines between the control signal generation and the switches in Fig. 5–25. After one modulation period T_m, the sequence repeats. Two alternative methods for generating the control signals are presented below in Sections 5.3.3.2 and 5.3.3.3.

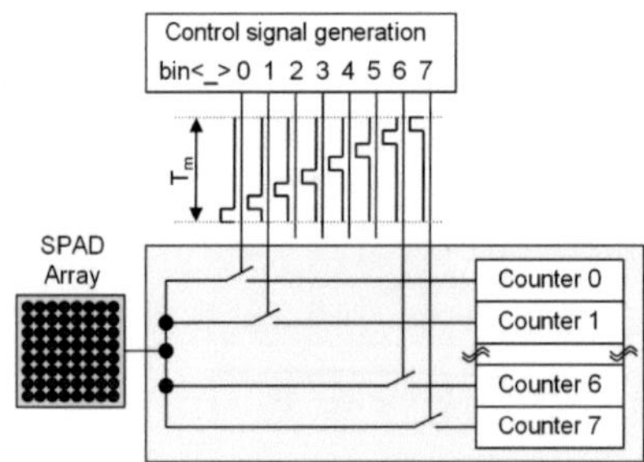

Fig. 5–25: Architecture for sorting SPAD pulses into counters. Switches controlled by control signals bin<0> to bin<7> sort incident SPAD events into corresponding counters.

5.3.3.2 Shift-Register-Based Binning

The RF output signal (CLK in Fig. 5–24) of the PLL provides the common source for generating the modulation of the laser diode and controlling the binning architecture. Fig. 5–26 (a) shows a shift-register-based binning architecture. A logic low at 'resetn' initializes the binning architecture. The outputs of the first seven flip-flops (FF) are set to logic low while the output of the last FF is set to logic high. During operation, 'resetn' is at logic high. With each clock signal at the 'CLK' input, the logic high shifts by one FF until the cycle repeats. The repetition frequency in this example is 1/8 of the clock frequency. Therefore, the clock frequency must be eight times higher than the desired modulation frequency. The outputs bin<0> to bin <7> are the control signals for the opening of the bins.

Fig. 5–26 (b) shows the logic levels of clock signal (CLK) and generated control signals bin<0> to bin<7> which define the sampling windows. As previously shown in Fig. 5–25, a SPAD event increments counter<0> when bin<0> is a logic high, counter<1> when bin<1> is a logic high, and so on.

Assuming that a modulation frequency of 1 GHz and 8 bins per modulation period are desired, the VCO of the PLL has to operate at a frequency of 8 GHz for generating the required clock signal CLK. The laser diode in turn receives the clock signal after a divider 1/8. Very accurate timing for the bins can be expected from this approach, as the clock signals for opening and closing the bins are derived from clock signals of a PLL that is locked to a stable reference oscillator. The individual pulses of the clock signal are equidistant time, which translates to equidistant bins of equal width. Hence, an almost ideal sampling process with sampling windows of finite width (see Appendix C) should be expected.

The implementation of a PLL operating at 8 GHz is highly desirable and certainly possible with state-of-the-art RF optimized fabrication processes, e.g., BiCMOS7RF with silicon-germanium option. However, SPADs are not yet available in these processes. As discussed previously, there are severe RF limitations in the imaging technology used for SPADs as a tradeoff for the required optical properties (see SPAD structure, Section 3.3.3). Hence, either the operating frequency must be reduced, with the corresponding increase of distance error due to photon-noise (see Eq. (5.28)), or alternative binning concepts have to be developed.

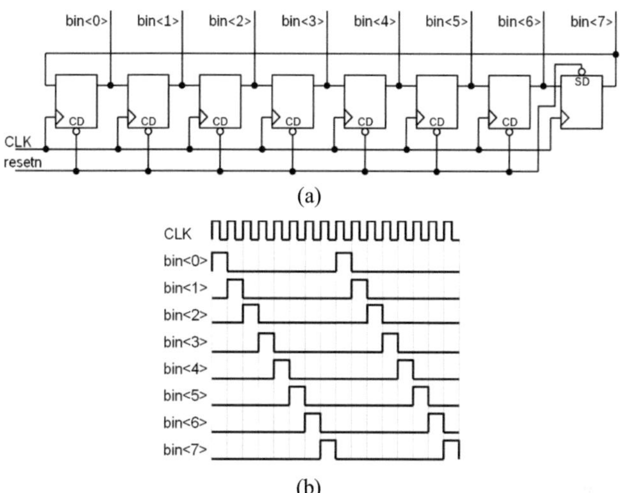

Fig. 5–26: Shift-register-based binning. (a) Binning architecture. A logic low at 'resetn' resets the binning architecture. The outputs of the first seven flip-flops (FF) are set to logic low, the output of the last FF is set to logic high. With each clock signal, the logic high shifts by one FF until the cycle repeats. The repetition frequency of the binning is 1/8 of the clock frequency. (b) Logic levels of clock signal (CLK) and generated control signals bin<0> to bin<7> which define the sampling windows. Two periods are shown.

5.3.3.3 Ring-VCO-Based Binning

A ring-VCO has favorable properties for a low-cost laser rangefinder. A ring-VCO features a lower power consumption, wide tunable frequency range and requires a smaller chip area than an LC-oscillator, which in turn reduces system cost [117]. A certain amount of phase noise can be tolerated, as long as it is small compared to SPAD jitter. Furthermore, a ring-VCO enables the implementation of a binning architecture that is completely different from the aforementioned shift-register binning.

Fig. 5–27 shows a schematic sketch of a ring-VCO comprising 4 differential delay stages. Each delay stage passes a logic level at its input on to the output with a certain delay. The delay can be adjusted via the delay control (delay ctrl). The sum of all delay stages defines the round-trip time of the loop and thus the VCO frequency. Each delay stage has a nominal delay of 1/8 of the period. A signal passing through the four delay stages on the upper path followed by passing through the four delay stages on the lower path gives the round-trip time. Partial signals $p1$ to $p4$ are the outputs of the first to fourth delay element on the upper path, whereas $\overline{p1}$ to $\overline{p4}$ are the inverted outputs of the first to fourth delay element on the lower path.

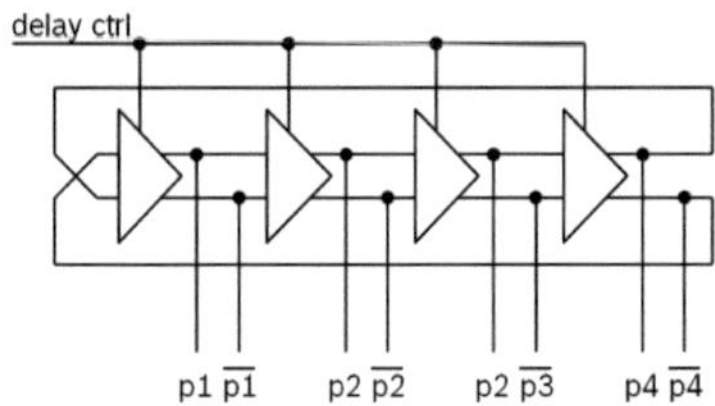

Fig. 5–27: Ring-VCO comprising four differential delay stages. A delay control signal (delay ctrl) controls the delay of the elements and thereby adjusts the frequency of the loop. At 1 GHz frequency, each delay stage has a nominal delay of 125 ps. Passing through the four delay stages on the upper path followed by the four delay stages on the lower path gives the round-trip period of 1 ns.

At a frequency of 1 GHz, each delay stage has a nominal delay of 125 ps. Passing through the four delay stages on the upper path followed by the four delay stages on the lower path gives the round-trip period of 1 ns. Fig. 5–28 (upper 4 rows) depicts partial signals p1 to p4 over time for two modulation periods. Any of these signals or their inverted signals can directly be used for modulating the laser diode at a frequency of 1 GHz. The control signals bin<0> to bin<7> are derived from the partial signals of the ring-VCO as shown in the lower 8 rows in Fig. 5–28. Favorably the partial signals at 1 GHz frequency are routed over the chip and logically combined where needed.

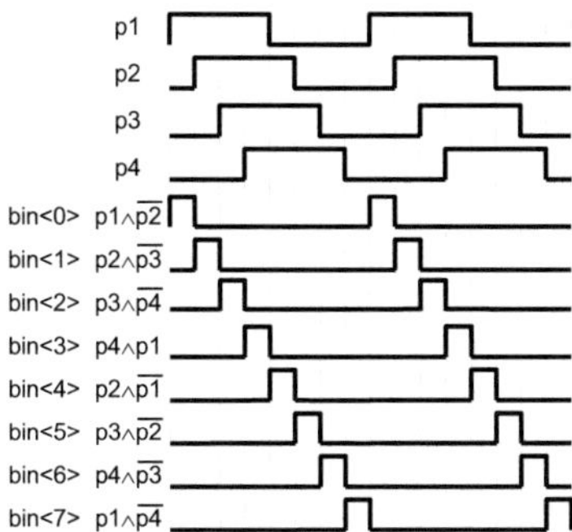

Fig. 5–28: Output of ring-VCO. Logic levels of outputs from the delay elements (upper 4 rows) and logic levels of control signals bin<0> to bin<7> generated from the output of the delay elements. Two periods are shown.

Fig. 5–29 presents a ring-VCO binning architecture with SPAD array and counters. Switches controlled by the control signals bin<0> to bin<7> forward SPAD pulses onto counters 0 to 7. Each pulse increments the associated counter. A microcontroller evaluates the phase of the received counter values and then calculates the target distance.

For the low-cost laser rangefinder targeted in this work, a ring-VCO-based binning architecture has been selected because of the lower power consumption, wide tunable frequency range and lower cost than an LC-oscillator. Thanks to access to the individual delay elements, multiple bins per period can be generated. Therefore, the VCO can be operated at a lower frequency compared to the shift-register-based binning of the previous section.

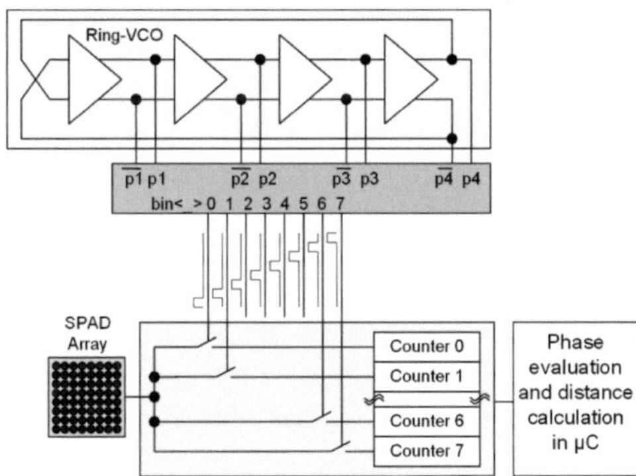

Fig. 5–29: Ring-VCO binning architecture with SPAD array and counters. Switches controlled by control signals bin<0> to bin<7> sort incident SPAD events into corresponding counters.

5.3.3.4 Non-Ideal Binning

The previous sections described idealized binning architectures. All N bins are equidistantly distributed in time, and each bin has an identical width of exactly $1/N$ of the modulation period. State-of-the-art distance measurement systems based on single-photon synchronous detection assume an idealized binning and do not apply any correction for unequal bin width or a SPAD sensitivity [39]. An example is the binning proposed by Niclass in [39], which operates at a modulation frequency of 30 MHz. At this moderate modulation frequency the mismatch between actual and ideal bin width is negligible compared to the overall bin width. In terms of the distance error due to photon noise and crosstalk, however, high modulation frequencies are mandatory for reaching millimeter precision (see performance estimation in Fig. 5–46). These high modulation frequencies give rise to new problems with respect to unequal bin widths.

A binning architecture derived from the individual delay stages of a ring VCO (see Section 5.3.3.3) relies on matching the delay of all stages. The mismatch of different delay stages is referred to as differential non-linearity (DNL) [118] (see Section 2.3.3). If one delay stage has a longer delay than the other stages this causes a wider width of the associated bin. Of course the PLL ensures that the round-trip time of all delay stages is locked, however, the PLL only controls the period time for all delay stages via a common control signal delay ctrl (see Fig. 5–27). Thus, the PLL controls the sum of the delay times but does not control each delay time of each delay element individually. PLLs with a ring-VCO comprising delay stages are also referred to as delay-locked-loops (DLL) [119].

The control signals for opening and closing the switches (see Fig. 5–29) have to propagate from the ring VCO to the switches. Parasitic capacitances and line resistance can decrease the edge slope and amplitude of the signals for opening and closing the bins and cause a pulse gap between bins.

There are both systematic and random binning non-idealities which disturb the binning and cause unequal bin widths. Systematic binning non-idealities depend on the specific environment of individual delay stages in the chip. For example, differences could be due to different parasitic capacitances and different line lengths. Systematic non-idealities are minimized in the design process. Random binning non-idealities are due to process, voltage and temperature (PVT) fluctuations that cause different delay times in the delay chain of the ring [118, 120]. Optimum matching of the elements of the ring is crucial [121].

An ideal binning and a non-ideal binning, a disturbed binning and a disturbed binning with gaps are exemplarily shown in Fig. 5–30 for one modulation period. Ideal bins (upper row) are equidistantly distributed over the modulation period and have equal bin widths. Non-ideal, disturbed bins (middle row) are generated from delay elements that suffer from different delay times. As the generated bins depend on the control signals provided by the delay stages, the bins are not equidistantly distributed over the period and have different bin widths. The lower row shows the disturbed bins further including gaps between the individual bins. SPAD pulses that fall into gaps are not counted.

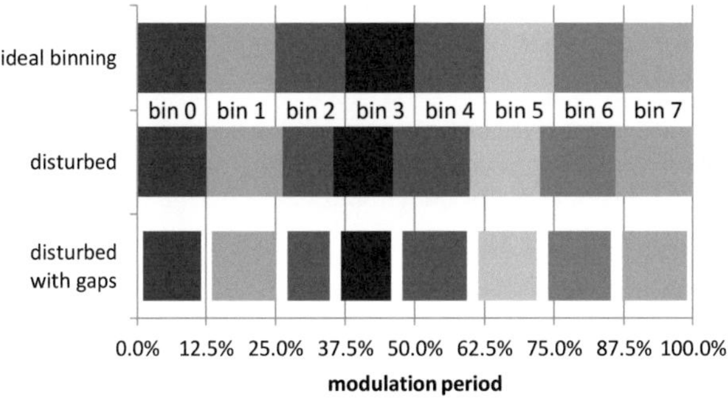

Fig. 5–30: Ideal and non-ideal disturbed binning. Each bin of the ideal binning with 8 bins covers 1/8 of the period (upper row). The bins of the disturbed binning are of unequal width and not equidistantly positioned, however, the PLL adjusts the overall time of the delay stages to one modulation period (middle row). Disturbed binning further including gaps between the individual bins (lower row).

The disturbed binning shown in Fig. 5–30 (middle row) is now used for sampling a received intensity modulated signal, as shown in Fig. 5–31. Sampling the received signal with an ideal binning is shown in Fig. 5–17 (page 151) for comparison. The number of photons detected in a bin depends on the bin width as previously explained for the ideal binning in Section 5.3.2.1. Thus, the counter values in Fig. 5–31 (c) not only depend on the received signal intensity Fig. 5–31 (a) but also on the disturbed bin width Fig. 5–31 (b).

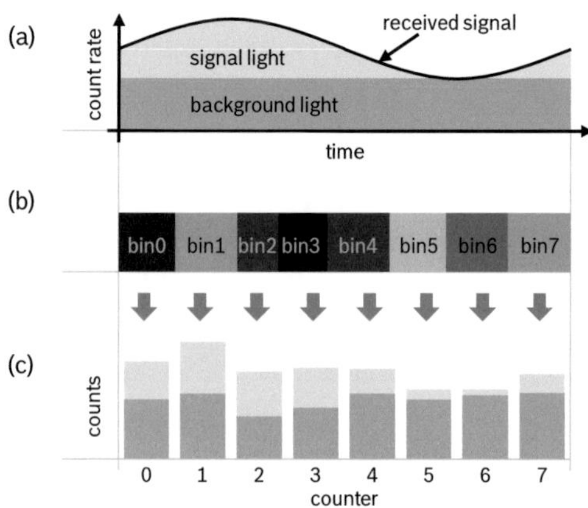

Fig. 5–31: Non-ideal disturbed binning. (a) Sinusoidal signal with offset from background light and SPAD dark counts. (b) Non-ideal disturbed bins of different widths that are non-equidistantly distributed over the modulation period. (c) The received signal is represented by a histogram of SPAD counts in each bin.

Fig. 5–32 shows a histogram of counts recorded with the disturbed binning. The received signal (solid black curve) comprises modulated signal light, having a sinusoidal intensity modulation, and an offset of constant background light plus SPAD dark counts. For illustration, light grey portions of the histogram bars represent counts from signal light, whereas dark grey portions represent counts from background light and dark counts. It should be noted that the output of the binning scheme is only one counter value per bin. Therefore, it is not possible to distinguish between counts from signal light, background light, or dark counts. Nonetheless, the counter values obtained with the non-ideal binning form the basis for phase evaluation of the fundamental modulation frequency as described for the ideal binning in Section 5.3.2.1 (see also Fig. 5–18 on page 152). For this disturbed binning, the phase of the reconstructed signal (dashed grey line) does not represent the phase of the received signal at all, but rather characterizes the non-ideal binning. In the shown example, the error phase from the non-ideal bin widths is dominant over the signal phase and must be corrected for.

As a marginal note, modulated pixels (see introductory chapter, Section 2.3.2) also perform a phase measurement to determine the target distance. The problem of unequal bin widths is similar to unequal sensitivity in modulated pixels, where the different CCD areas can have different sensitivity [30, 122]. Similar to a wider bin width, a pixel with higher sensitivity detects more light. Different architectures and geometric layouts for such modulated pixels are proposed in [123].

Non-idealities of the binning not only affect the bin width, but also affect the position of the bin center (see Section 2.3.3, Fig. 2–14). The bins are not equidistantly distributed in time. Equidistant samples, however, are a prerequisite for the discrete Fourier transform to work properly. This effect further disturbs the measurement.

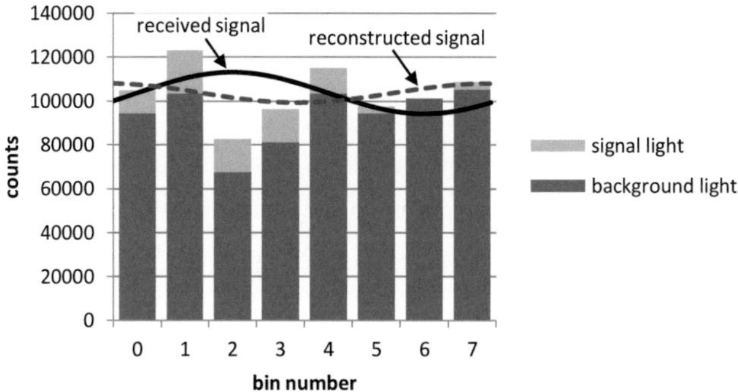

Fig. 5–32: Histogram of counts detected with non-ideal bins of different widths. The received signal (solid black curve) is a sinusoidal signal with offset from background light and SPAD dark counts. The height of the bars is defined by the signal intensity and the bin width. The reconstructed signal (dashed grey curve) is calculated from the Fourier transform of the measured counts.

5.3.4 Code Density Test

A code density test (CDT) aims at measuring the different bin widths for applying a calibration, which equalizes these different bin widths. The name "code density test" originates from the characterization of analog-to-digital converters [124].

The measurement of bin widths is straightforward and can easily be implemented in a low-cost device. For simplicity, in the following description the term 'background light' covers both background light and SPAD dark counts. The modulated signal light is turned off such that the SPAD is illuminated with background light only, which is constant on the scale of measurement. (Strictly speaking this requirement can be relaxed to background light uncorrelated with the binning. On average a constant number of photons should be incident on the SPAD. For example a light bulb or fluorescent lamp connected to the AC power grid works well, so there is no need for an extra costly light source for calibration.) Counters record all photons falling into the respective bins. Hence, under the condition of constant illumination, the counter values represent the bin width.

A relative bin width $w_{\mathrm{rel},k}$ can be calculated for each bin from the measured counter values $x_{k,\mathrm{BL}}$ by

$$
w_{\mathrm{rel},k} = \frac{x_{k,\mathrm{BL}}}{\frac{1}{8}\sum_{l=0}^{7} x_{l,\mathrm{BL}}} \quad , \quad c_k = \frac{1}{w_{\mathrm{rel},k}} \; . \tag{5.29}
$$

It should be noted that the relative bin width does not take pulse gaps (see Fig. 5–30) into account. Thus the relative bin width only measures how much of the total opening time of all bins is allotted to each bin. An ideal bin has a relative bin width of 100 %. In contrast to the relative bin width, the absolute bin width also takes pulse gaps into account. To determine the absolute bin width, another counter is required that continuously measures incident counts over the entire period. That counter value can then replace the sum in the denominator of Eq. (5.29). To compensate for bin width errors however, the absolute bin widths are not required as we are interested in equalizing a bin width difference.

The code density test (Eq. (5.29)) is also not capable of actually measuring the bin position. As the background light is constant with time, the measured counts are invariant to the temporal bin position. The bin position depends on the bin width to a certain extent. For example a first large bin indicates a long delay element of the delay line, thus the following bins open later which in turn shifts the bin centers of the subsequent bins to a later time (see Section 2.3.3.1). Furthermore, the gate signals for opening and closing individual bins propagate over the chip, and thus experience further delays which can differ from bin to bin. An accurate measurement of the bin width is still possible with constant illumination. However, this does not give precise knowledge about the actual bin position. A measurement of actual bin position requires a timing reference, which is not feasible in a handheld low-cost device. While the bin width is a major source of error that can completely obscure the actual signal phase, as shown in Fig. 5–32, a deviating bin position will 'only' cause a residual minor phase error. Overall, the bin width is the dominant source of error that must be corrected for.

Both hardware correction and software correction are possible to mitigate binning non-idealities.

Hardware correction

Hardware correction for the binning requires individual adjustments for the delay elements of the ring-VCO to equal delay times. Baronti et al. present an all-digital adjustable DLL [125]. They adjust the delay of each individual delay element of the ring by adding digitally controlled shunt capacitors to the output of each individual delay element. After hardware-calibration, the maximum differential non-linearity (DNL) error is 3.5 ps corresponding to a 3 % bin width error for a 125 ps wide bin. Mota et al. implemented an 8-tap DLL operating at 2.5 GHz round-trip frequency in 0.7μm CMOS [126]. After hardware-calibration, the standard deviation of the bin width is still 21.5 ps.

While hardware-calibration schemes have been shown to reduce the DNL error, the compensation is not perfect yet. A bin width error of 3.5 ps is a remarkable precision, however, a problem persists in a laser rangefinder when measuring a weak signal under strong background light conditions. Assuming a signal modulation depth of 10 %, the error phasor magnitude from a bin width mismatch of only 3% is of the same order as the signal phasor, which can cause several centimeters distance error (depending on the phase relation of the signal phasor with respect to the error phasor, Fig. 5–2). Moreover, the on-chip hardware-calibration increases the complexity of the system, increases the required silicon area and hence increases the system cost.

Software correction

Software correction of bin widths is easy and cost effective. The measured counter values are simply divided by the bin widths determined in a previous bin width measurement with constant background light. A simple division can of course be computed by a low-cost microcontroller which is already available in the laser rangefinder.

As previously shown in Eq. (5.16), the phase is computed from the counter values for bins 0 to 7 by means of a discrete Fourier transform (DFT), but now the additional correction factors $c_{0...7}$ from Eq. (5.29) are added,

$$X_l = \frac{1}{N}\sum_{k=0}^{N-1} c_k x_k e^{-j2\pi\frac{kl}{N}} \, , \; N = 8 \, , \; l = 0,1...(N-1) \, . \tag{5.30}$$

The correction factors $c_{0...7}$ are the inverse of the (relative) bin widths $w_{0...7}$ measured with non-modulated background light only. The quantities $x_{0...7}$ are the associated counter values measured with both background light and modulated signal light received from the target. The calculated Fourier components are denoted $X_{0...7}$. The target distance is determined from the phase of X_1.

Fig. 5–33 illustrates the steps required for CDT-correction. Step (a) shows a histogram of counter values measured with the corresponding bins for constant background light only. Step (b) presents the counter values for both background light (dark grey bars) and modulated signal light (light grey bars). As previously shown in Fig. 5–32, the measured phase at the fundamental modulation frequency does not correspond to the actually received signal phase. In step (c) the measured counter values from step (b) are divided by the relative bin widths from step (a) to compensate the counts x_k with a factor c_k (Eq. (5.29)) for bin width non-idealities.

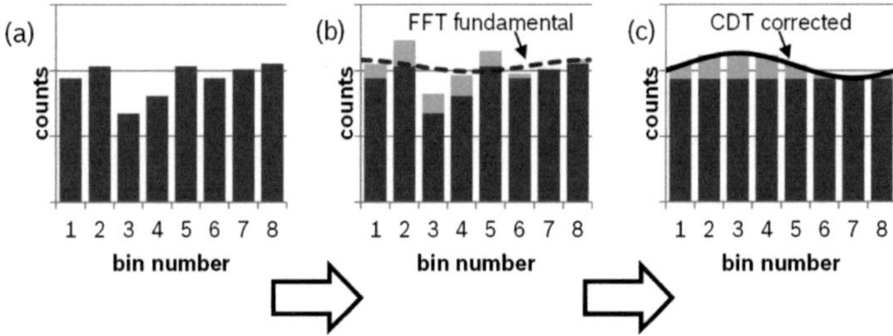

Fig. 5–33: Steps of CDT-correction. (a) Determination of bin widths from counter values measured with homogeneously distributed background light. (b) Measurement of signal and background light with non-ideal binning. (c) Divide measured counter values by the corresponding bin width (CDT-correction) and determine the phase at the fundamental frequency.

As shown in Fig. 5–33 (c), ideal CDT software correction perfectly equalizes the bin width error for background light. Hence, this error source, which dominates measurements at strong background light and weak signal modulation depth, can be fully eliminated. However, this error compensation is not complete for the signal light.

While the dominant error from background light can be compensated for, there are two limitations to this software correction of binning non-idealities. A non-equidistant bin position is not corrected and still causes a residual error. Furthermore, each bin – considered as a rectangular sampling window – has a different filter characteristic in frequency domain (see Section 5.3.2.2). Considering a modulation frequency of 1 GHz, each of the eight bins should ideally span 125 ps. A bin that is 20% too small, i.e., 100 ps wide, transmits 98.4% of the signal at the fundamental modulation frequency, whereas a bin that is 20% too large, i.e., 150 ps wide, transmits only 96.3%. For a non-modulated signal, i.e., for background light, the filter characteristic is unaffected and has a transmission of 100 %.

Furthermore, it should be highlighted that a calibration measurement is mandatory to determine the bin width for CDT-correction. The accuracy of bin width correction can only be as good as the accuracy of the calibration, hence the accuracy of the bin width measurement prior to correction is crucial.

The bin width is determined from the counts due to background light falling into each bin. Hence, assuming a large average number $\overline{N}$ of Poisson distributed events, each bin width is known with a relative accuracy $1/\sqrt{\overline{N}}$ (see Eq. (3.13)). An additional light source increases $\overline{N}$ and can thus be employed to improve the accuracy of the calibration. Alternatively, as a cost-effective solution, the laser of the rangefinder simply be operated in DC mode during calibration [127] to increase $\overline{N}$.

A calibration via a code density test and software correction is not desired before each distance measurement because of the mandatory calibration. CDT would be feasible if all fabricated ASICs exhibited identical behavior or if a one-time calibration in manufacturing were sufficient. As will be shown experimentally in Chapter 6, the binning of each ASIC has to be calibrated individually. Even for one ASIC a calibration is needed at each measurement frequency (Section 2.2.3.3). Furthermore, the bin width changes with ambient conditions.

Results from a Monte Carlo simulation for a laser rangefinder with 8-tap binning using CDT-correction will be shown exemplarily in the performance estimation in Section 5.4.1.

5.3.5 Bin Homogenization

The goal of the 'bin homogenization' presented in this work is a compensation scheme that corrects for bin width errors (i.e. sampling windows of unequal width) and does not require a calibration measurement prior to the actual distance measurement. Most intriguingly, different bin widths can be compensated for without even knowing the respective bin widths. Pending patent applications by the author are disclosed [96] or about to be published [97].

5.3.5.1 Manipulating the Periodicity

The core idea of bin homogenization is to manipulate the periodicity of the histogram of SPAD counts.

As explained before, the binning derived from the ring-VCO suffers from non-idealities (Section 5.3.3.4). The received signal, comprising the modulated signal light and constant background light, is sampled with the non-ideal binning, and a histogram of SPAD counts falling into the different bins (sampling windows) is obtained (Fig. 5–31 (c)). For constant background light, the different counter values due to background light (dark grey portions in Fig. 5–31 (c)) correspond to the bin widths. These different counter values can be misinterpreted as a modulation of the received signal. Especially in cases of strong background light this pseudo-modulation due to background light, which is detected with the non-ideal binning, can exceed the desired modulation due to signal light (Fig. 5–32).

The target distance is determined based on the phase Φ_{Rx} of the received signal at the fundamental modulation frequency, which is determined from the counter values of the histogram (Eq. (5.16), Eq. (5.17)). It is thus an objective to eliminate the impact of background light on the fundamental modulation frequency.

As a solution to this problem, we suggest to manipulate the effective periodicity of the binning. As described in Appendix C, a periodic signal has a frequency-discrete spectrum, wherein the frequency components are located at integer multiples of the inverse of the period, $1/T$, which in our case corresponds to the fundamental modulation frequency f_m. This situation is similar to sampling the constant background light with a periodic sequence of bins. The repetition frequency of the disturbed bins is denoted by f_{rep}. The disturbed binning thus has a frequency-discrete spectrum, wherein the frequency components are located at integer multiples of the repetition frequency f_{rep}. So far, the repetition frequency of the bins corresponds to the modulation frequency $f_{\text{rep}} = f_m$. Therefore, disturbances of the binning at the repetition frequency f_{rep} can coincide with and disturb the fundamental modulation frequency f_m.

Thus, to avoid such disturbances, the repetition frequency of the binning should be manipulated such that it differs from the fundamental modulation frequency. Since the system cost should remain very low, it would be desirable to derive both the modulation frequency as well as the binning from the same PLL. To save chip area, also the number of counters should be kept low. Furthermore, an easy evaluation which involves processing few counter values and few calculations is desired.

One theoretical option would be to reduce the number of bins by half, e.g. from 8 to 4 bins, and to operate the binning (and also the PLL from which the binning is derived) at twice the modulation frequency $f_{\text{rep}} = 2f_m$. Therefore, disturbances of this modified binning are located at integer multiples of twice the modulation frequency and do not disturb the fundamental modulation frequency f_m.

However, the operation frequency of the SPAD fabrication process is limited such that the frequency of the PLL cannot simply be doubled. Instead, the modulation frequency would have to be reduced. This is not desired, since the distance accuracy scales with the modulation frequency (Eq. (2.10)). Moreover, one would have to distinguish whether as SPAD event has been detected in the first half or the second half of the modulation period. Thus, the binning would have to be followed by a second stage of sorting SPAD events into counters.

Instead of significantly modifying the binning, we suggest to perform two subsequent sub-measurements, wherein the position of the binning is changed with respect to the received signal. The concept is sketched in Fig. 5–34. During a first sub-measurement (sub1) SPAD counts falling into bin 0 increase counter 0, SPAD counts falling into bin 1 increase counter 1, and so on. During a second sub-measurement (sub2), SPAD counts falling into bin 0 increase counter 4, SPAD counts falling into bin 1 increase counter 5, SPAD counts falling into bin 4 increase counter 0, and so on. The counts of both sub-measurements subsequently increment the counters. No additional counters are required, the existing counters are reassigned.

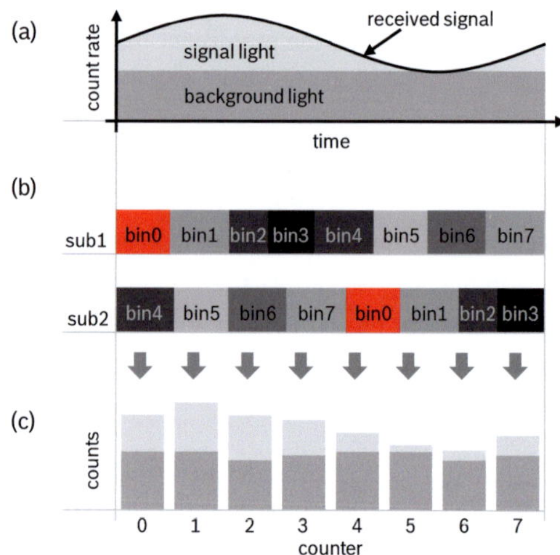

Fig. 5–34: Bin homogenization with two sub-measurements. (a) Sinusoidal signal with offset from background light and SPAD dark counts. (b) Two sub-measurements (sub1, sub2) with non-ideal disturbed bins of different widths that are non-equidistantly distributed over the modulation period. (c) The sum of the counts from both sub-measurements builds a histogram.

With this new concept, the fixed connection between counter and bin is broken up. It should be highlighted again that the bin defines the sampling window. Any time delays that are introduced after the bin, i.e., after sampling the received signal, do not affect the sampling process. Even if the connection between bin 0 and counter 0 during the first sub-measurement shows a different delay time than the connection between bin 0 and counter 4 during the second sub-measurement, this does not impact the sampling process since the routing from the bin to the counter is not on the time-critical path.

The two sub-measurements as shown in Fig. 5–34 effectively change the periodicity of the sampling process. Fig. 5–35 (a) illustrates that bins 0 and 4 contribute to counter 0 and that bins 0 and 4 contribute to counter 4. Every 4th counter is measured with the same two bins. Thus, on average every 4th counter sees the same average bin width. This average bin width is referred to the effective bin width,

$$w_{\text{eff},k} = \frac{1}{2}\left(w_{\text{rel},k} + w_{\text{rel},k+N/2}\right) \quad \text{for} \quad k = 0,1,...N/2\text{-}1$$

$$w_{\text{eff},k} = w_{\text{eff},k-N/2} \qquad\qquad \text{for} \quad k = N/2...N. \tag{5.31}$$

The normalized counts in Fig. 5–35 (a) illustrate the effective bin width. The repetition frequency is effectively doubled to $f_{\text{rep}} = 2f_m$.

A discrete spectrogram calculated from the counter values is shown in Fig. 5–35 (b) for one single sub-measurement (sub 1) representative of the non-ideal binning (solid grey line) and for the combination of sub-measurements 1 and 2 representative (dashed black line). For a period of 1 ns the fundamental modulation frequency for distance measurement is 1 GHz. Without bin homogenization the non-ideal binning has an impact on the integer multiples of the modulation frequency. In particular the disturbance on the fundamental modulation frequency deteriorates a distance measurement. However, with the two sub-measurements as a correction, the non-ideal binning only impacts integer multiples of twice the modulation frequency and effectively eliminates the impact on the fundamental modulation frequency.

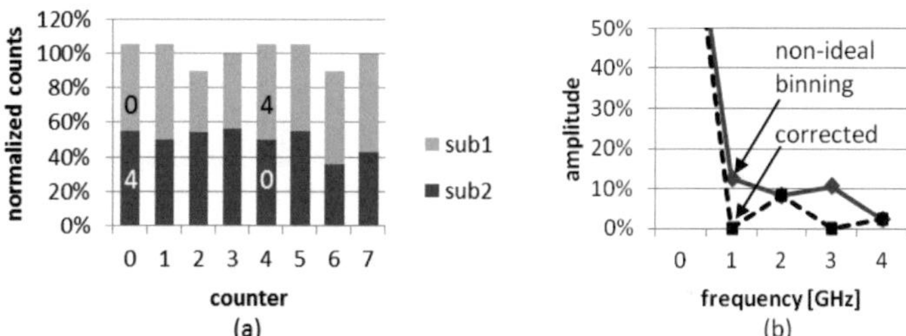

Fig. 5–35: Bin homogenization with two sub-measurements (sub1 and sub2). (a) Counter values for sub-measurements 1 and 2. Numbers on the bars indicate the bin with which said bar is measured. Every 4th counter value is measured with the same two bins, e.g., bin 0 and 4. (b) Discrete spectrum calculated from counter values assuming 1 ns period and 8 bins, with and without correction. Amplitude normalized to 100 % DC level.

To formalize the effect of bin homogenization, let $X_{0...7}$ be the calculated Fourier coefficients, $w_{\text{rel},k}$ the relative bin width for non-ideal bin k (Eq. (5.29)), $\bar{n}_{BL}$ be the average count rate from background light and t_{meas} the measurement time. The counter values are given by $x_k = \bar{n}_{BL} t_{\text{meas}} w_{\text{rel},k}$, wherein the average count rate $\bar{n}_{BL}$ and the measurement time t_{meas} are the same for all bins. The Fourier coefficients are calculate as previously described in Eq. (5.16) by

$$X_l = \frac{\overline{n}_{\mathrm{BL}} t_{\mathrm{meas}}}{N} \sum_{k=0}^{N-1} w_{\mathrm{rel},k} e^{-j2\pi\frac{kl}{N}} \ , \ N = 8 \ , \ k = 0,1,...N-1.$$

(5.32)

Because the same effective bin width (Eq. (5.31)) repeats every $N/2$, i.e., $w_{\mathrm{rel},k} = w_{\mathrm{rel},k-N/2}$, this can be written as

$$X_l = \frac{\overline{n}_{\mathrm{BL}} t_{\mathrm{meas}}}{N} \left(\sum_{k=0}^{N/2-1} w_{\mathrm{rel},k} e^{-j2\pi\frac{kl}{N}} + \frac{1}{N} \sum_{k=N/2}^{N-1} w_{\mathrm{rel},k} e^{-j2\pi\frac{kl}{N}} \right)$$

substituting $k = k' - N/2$ in the right - hand side sum

$$= \frac{\overline{n}_{\mathrm{BL}} t_{\mathrm{meas}}}{N} \left(\sum_{k=0}^{N/2-1} w_{\mathrm{rel},k} e^{-j2\pi\frac{kl}{N}} + \frac{1}{N} \sum_{k'=0}^{N/2-1} w_{\mathrm{rel},k'} e^{-j2\pi\frac{k'l}{N}+j2\pi\frac{l}{2}} \right)$$

(5.33)

$$= \frac{\overline{n}_{\mathrm{BL}} t_{\mathrm{meas}}}{N} \left(\sum_{k=0}^{N/2-1} w_{\mathrm{rel},k} \overline{n}_{\mathrm{BL}} e^{-j2\pi\frac{kl}{N}} + (-1)^l \frac{1}{N} \sum_{k'=0}^{N/2-1} w_{\mathrm{rel},k'} \overline{n}_{\mathrm{BL}} e^{-j2\pi\frac{k'l}{N}} \right)$$

$$= \frac{\overline{n}_{\mathrm{BL}} t_{\mathrm{meas}}}{N} \left(\sum_{k=0}^{N/2-1} \left[w_{\mathrm{rel},k} + (-1)^l w_{\mathrm{rel},k} \right] \overline{n}_{\mathrm{BL}} e^{-j2\pi\frac{kl}{N}} \right).$$

It follows from the term $\left[w_{\mathrm{rel},k} + (-1)^l w_{\mathrm{rel},k} \right]$ in Eq. (5.33) that only even Fourier coefficients are present for background light, while there is no impact of background light on odd Fourier coefficients, since

$$w_{\mathrm{rel},k} + (-1)^l w_{\mathrm{rel},k} = \begin{cases} 0 & \text{for odd } l \\ 2w_{\mathrm{rel},k} & \text{for even } l \end{cases}$$

(5.34)

Hence, with the proposed scheme, odd harmonics vanish. This includes any impact of background light on the fundamental modulation frequency. Even harmonics are still present but are not evaluated for distance measurement. The impact of background light on the fundamental modulation frequency cancels perfectly. A calibration measurement prior to the actual distance measurement is not required.

As a limitation, it should be noted that the effective bin widths can still be different. The received signal is thus still sampled with finite sampling windows of different widths (Appendix C), which sampling windows can have different filter characteristics on the modulated signal (Section 5.3.2.2). Furthermore, as with the code density test correction by software in the previous section (Section 5.3.4), the non-equidistant positions of the bins are not corrected for.

There are two prerequisites for bin homogenization. Firstly, an identical measurement time per sub-measurement must be ensured for equal integration time and hence same counter values from background light. Secondly, the lighting and receiving conditions must not change between the sub-measurements. It should be noted that the two sub-measurements can be repeated over and over again. Thus multiple measurement cycles can be repeated, each with a short measurement time. Hence, the lighting conditions have to be stable at least during each short measurement time.

5.3.5.2 Homogenization

By increasing the number of sub-measurements from two sub-measurements (Fig. 5–34) to eight sub-measurements (Fig. 5–36), the background light can be completely homogeneously distributed over all counters. During a first sub-measurement (sub0) SPAD counts falling into bin 0 increase counter 0, SPAD counts falling into bin 1 increase counter 1, and so on. During a second sub-measurement (sub1), SPAD counts falling into bin 0 increase counter 1, SPAD counts falling into bin 1 increase counter 2, and so on, and SPAD counts falling into bin 7 increase counter 0. This scheme is continued for all sub-measurements. In the end all eight bins once contributed counts to each of the eight counters.

With bin homogenization with eight sub-measurements, the effective bin width is the same for all eight bins

$$w_{\mathrm{eff},k} = \frac{1}{N} \sum_{l=0}^{N-1} w_{\mathrm{rel},l} \quad \text{for} \quad k = 0,1,...N-1. \tag{5.35}$$

Hence, even though the bin widths are non-ideal, the background light is homogeneously distributed over all counters. Knowledge about bin widths is not required for compensation.

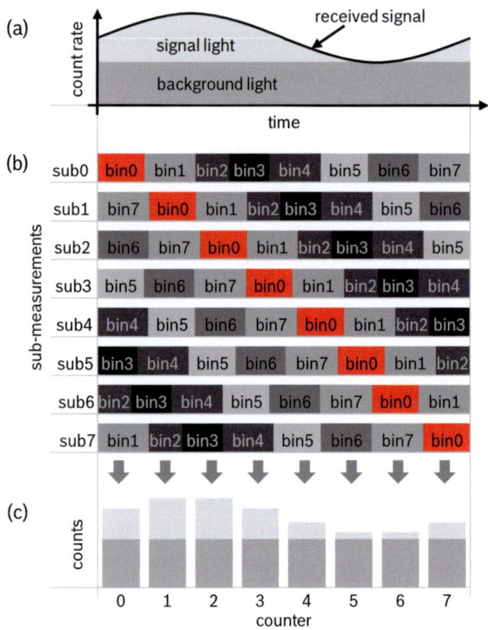

Fig. 5–36: Bin homogenization with eight sub-measurements. (a) Sinusoidal signal with offset from background light and SPAD dark counts. (b) Eight sub-measurements (sub0…sub7) with non-ideal disturbed bins of different widths that are non-equidistantly distributed over the modulation period. (c) The sum of the counts from all sub-measurements builds a histogram.

An example for the homogenization effect for background light is given in Fig. 5–37. In Fig. 5–37 (a) each bar of the histogram represents the counts sampled with one bin (normalized to 100 % mean value). Each bin is derived from the disturbed binning and has a different width. With bin width homogenization as shown in Fig. 5–37 (b), each bar represents the counts of one counter. Each counter receives SPAD events from all bins as exemplarily shown in red for bin 0.

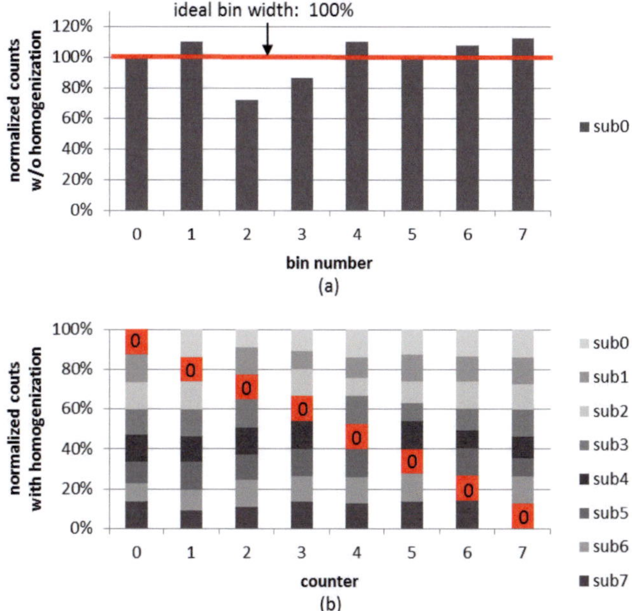

Fig. 5–37: Theoretical homogenization of background light. (a) Without bin width homogenization. (b) With bin width homogenization. Each row corresponds to a sub-measurement (sub0…sub7). The red elements mark the portion in each counter that is measured with bin 0.

It should be noted that bin homogenization mitigates the effects of different bin widths. However, bin homogenization does not distribute the bins equidistantly over the modulation period (Fig. 5–36). A non-ideal bin position does not affect the homogenization of background light because the background light is constant over time. However, the sampling of the modulated signal light can still cause a residual disturbance. This error contribution is included in the Monte Carlo simulation in system performance estimation in Section 5.4 below.

5.3.5.3 Implementation

The problem when implementing this scheme is how to change the binning between different sub-measurements (see Fig. 5–36) without changing the bin width ever so slightly. Even though a non-perfect temporal shift does not impact the number of measured counts from constant background light, any change in bin width has detrimental effects. Hence, it is of utmost importance that the physical connection between ring-VCO and binning is maintained without any additional elements for manipulating the binning.

The solution is to perform the mathematical equivalent to shifting the position of the bin in the receive path, i.e., shifting in time the modulated signal of the transmitter. Fig. 5–38 (a) illustrates a time shift of the modulated signal by the temporal width of bin 0.

The sub-measurement with the unchanged binning, Fig. 5–38 (b), corresponds to sub-measurement sub7 of Fig. 5–36. The periodic repetition of the binning is indicated by repeating bin0 at the far right side of Fig. 5–38 (b).

The temporal shift of the receiver is undone at the connection between bins and counters. The assignment of the counters to the unchanged binning, Fig. 5–38 (c), again corresponds to the sub-measurement sub7 of Fig. 5–36.

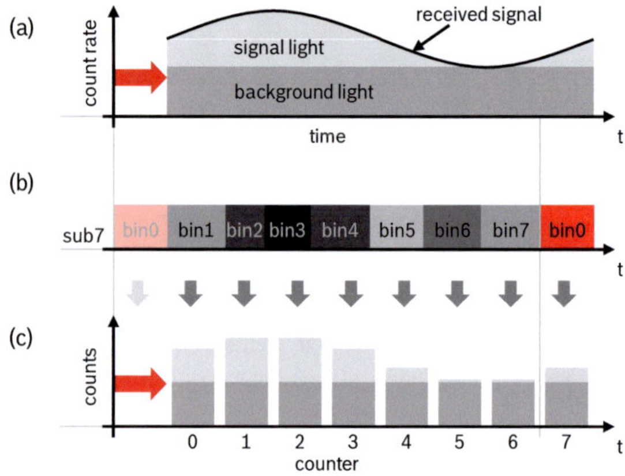

Fig. 5–38: Implementation of binning without changing the binning, exemplary sub-measurement corresponding to sub7 in Fig. 5–36. (a) The temporal position of the transmitted signal and thus the position of the received signal is shifted by the temporal width of bin 0. (b) The binning can remain unchanged between different sub-measurements. (c) The temporal shift of the receiver is undone at the connection between bins and counters, such that the assignment of the counters to the unchanged binning again corresponds to the sub-measurement sub7 in Fig. 5–36.

The required temporal shift between the sub-measurements depends on the bin width. On average, the modulated signal has to be shifted by an eighth of the modulation period, $T/8$, for eight bins. Since the temporal shift is coupled to the period duration, this shift can also be referred to as a phase shift with respect to the modulation period. The average phase shift between sub-measurements is thus $2\pi/8$. The corresponding structural element of the laser rangefinder ASIC can be referred to as a phase shifter. A Tx phase shifter causes the shift of the output signal of the modulator, whereas an Rx phase shifter causes the opposed shift between binning and counters at the receiver. A block diagram of the hardware implementation in the laser rangefinder ASIC CF340 MPW2011 will be shown in Section 6.2.1, Fig. 6–16 on page 215.

5.3.6 Electrical Crosstalk

Electrical crosstalk occurs in addition to optical crosstalk (see Section 5.2.4.3). The laser modulator is the primary source of crosstalk, because of the 'high' current (some milliamperes) for modulating the laser diode. Firstly, crosstalk from the laser modulator to the SPAD bias affects the photon-detection efficiency (PDE) of the SPAD. Secondly, crosstalk from the laser modulator to the ring oscillator affects the binning due to threshold shifts of ring oscillator and switches. As sketched in Fig. 5–39, both influences give a combined crosstalk error, i.e., different counter values are expected if the laser modulator is switched on or off.

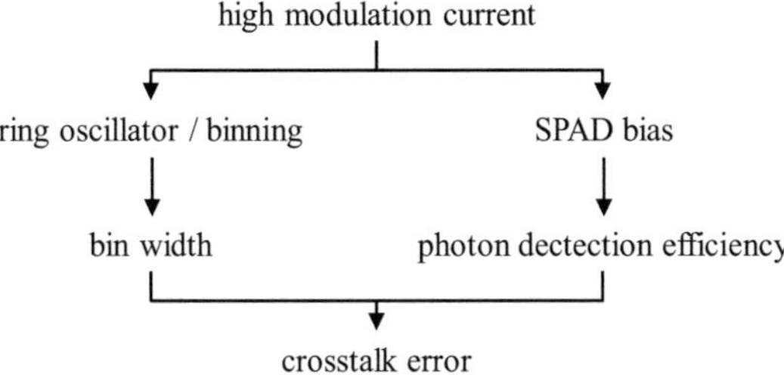

Fig. 5–39: Electrical crosstalk. The laser modulator is the primary source of crosstalk, because of the high current for modulating the laser diode. Crosstalk from the laser modulator to the SPAD bias affects the photon-detection efficiency (PDE) of the SPAD. Crosstalk from the laser modulator to the ring oscillator affects the binning due to threshold shifts of ring oscillator and switches. Both influences give a combined crosstalk error.

The current modulation of the laser modulator periodically alters the voltage levels within the ASIC. As experimentally shown in the SPAD characterization in Chapter 4, the PDE of the SPAD depends on the bias voltage (see Fig. 4–11, on page 100). Analogous to modulating the gain of an APD in a conventional APD-based laser rangefinder, a periodic variation of the bias voltage modulates the PDE of the SPAD.

Furthermore, any change in supply voltage can impact the switching behavior/decision thresholds of elements such as the delay stages of the ring-

VCO, binning logic or control signals for the switches of the binning process. Hence, the bin width and bin position are affected by crosstalk.

One way to reduce the impact of crosstalk is the use of different voltage domains within the ASIC for sensitive circuit elements such as the binning. However, decoupling of voltage domains and filtering is limited within one single chip.

Bin homogenization suffers particularly from electrical crosstalk from the laser modulator onto the binning because bin homogenization requires that bin widths do not change between sub-measurements. If the bin width changes between sub-measurements, a background-light-dependent error can be expected.

Electrical crosstalk can be mitigated to a certain extent when measuring distances with CDT-correction (see Section 5.3.4). Bin widths for CDT-correction are usually determined in a calibration measurement with the laser driver turned off, i.e., before the actual distance measurement and in absence of crosstalk from the laser modulator. Instead of performing the calibration measurement with the on-chip laser modulator turned off, we propose to perform said calibration measurement with the laser modulator turned on – however with the DC current of the laser diode set such that the laser diode is modulated below lasing threshold (Bosch patent application [128]). Any electrical crosstalk, no matter if the bin width or the PDE is altered, is thereby already included in the correction factors for CDT-correction. This method is referred to as sub-threshold calibration [128].

5.3.7 Delay Drift

In the transmit path, the temperature behavior of the laser diode has already been identified as the dominant contributor to a drift of the measured phase (see Section 5.2.3.1). A laser rangefinder typically features a target and a reference path to compensate for drifts of the transmitter and the receiver. The actual target distance is calculated from the difference between target and reference path lengths.

However, also different SPADs, i.e., one target and one reference SPAD or different pixels of a SPAD array, can show a different temporal characteristic. A constant temporal offset, i.e., a time difference between outputs of two SPADs, can be calibrated once in the manufacturing process. However, delay drifts occur because of temperature and voltage fluctuations as well as aging of the device. The difference in the temporal behavior of target and reference SPAD depends (a) on the time when a photon hits each of the SPADs until an output pulse is released and (b) on the time from an individual pixel output until the leading edge is recognized by the timing circuitry. At an angular modulation frequency $\omega_m = 2\pi f_m$, a delay drift causes a phase error of

$$\Delta\varphi_{\text{drift}} = \omega_m \tau_{\text{drift}} \qquad (5.36)$$

and a distance measurement error of

$$\Delta L = \frac{1}{2} c \tau_{\text{drift}} \, . \qquad (5.37)$$

In general, same types of components, e.g. SPADs or output buffers, within one chip will show the same overall temporal behavior and delay drift in the same direction. The difference in delay drift between target and reference SPAD can thus be reduced by selecting the same components and by a strictly symmetrical chip layout for target and reference path as sketched in Fig. 5–40. An absolute delay drift thereby reduces to a differential timing drift, i.e. a drift difference, between target and reference SPAD [20].

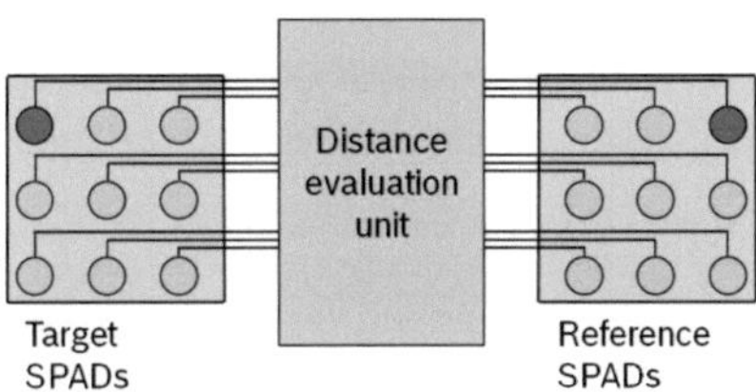

Fig. 5–40: Target and reference array with distance evaluation unit. SPADs at symmetrical positions of the two arrays are evaluated.

The actual values of the drift depend on the semiconductor technology being used. A detailed simulation has been performed by our colleagues at Bosch Automotive Electronics. The result is included in the system performance estimation in Section 5.4.

It should be noted that the problem of delay drift does not exist for laser rangefinders using a mechanical reference flap to switch between target and reference path, because the same detector is used for target and reference measurement. The proposed SPAD-based laser rangefinder, however, should not rely on a costly mechanical reference flap but feature independent detectors for target and reference measurement integrated on-chip.

5.4 System Performance Estimation

The following section presents the results of a system performance simulation based on calculated and estimated values from the previous sections. Please refer to Fig. 5–1 for an overview graph of the system performance estimation procedure. Remarks on the Monte Carlo simulation of the binning with CDT-correction and bin homogenization are provided first before presenting the estimated distance measurement error that can be achieved with the proposed laser rangefinder using the previously discussed low-cost photon counter (SPAD).

5.4.1 Monte Carlo Simulation of Binning and Correction

This section presents simulation results for a Monte Carlo simulation of binning and correction schemes.

The actual bin width and bin position, and thereby the influence of the binning on the reconstructed signal, depends on the opening and closing of the bins derived from the non-ideal delay elements. The actual matching properties of the delay elements depend on the semiconductor technology and can be provided by the technology supplier [129]. The delay time of the non-ideal delay elements is assumed to have a Gaussian distribution. The variance represents the time tolerance.

We used this information to set up a Monte Carlo simulation that models randomized opening and closing control signals for the binning. Each such a binning samples an input signal, and is responsible for the phase error between the phase of an input signal and the phase of the reconstructed signal. The distance error can be calculated from the phase error using Eq. (2.10). The input signal in turn is derived from the information about the transmitter (Section 5.2.3), the receiver (Section 5.2.4), and SPAD properties. Expected count rates (Fig. 5–16) and modulation depth of the signal (Fig. 5–22) have been discussed earlier in this chapter. The input signal is assumed free of noise. A diagram of the performance estimation concept has been shown in Fig. 5–1.

The error between input signal and reconstructed signal is evaluated for input signal phases from 0 to 2π. The results are saved for each phase position of the input signal, and the entire process is repeated for N iterations. A flow chart of the Monte Carlo simulation is shown in Fig. 5–41. During the step of correcting counter values, different subroutines b) code density test (see Section 5.3.4) and c) bin homogenization (see Section 5.3.5) can be employed for correction.

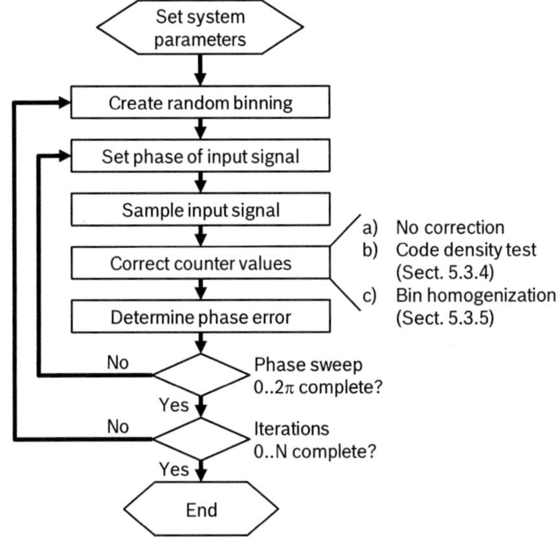

Fig. 5–41: Flow chart of Monte Carlo simulation to determine measurement error from non-ideal binning.

The rows in Fig. 5–42 represent high, medium and low modulation depth of the received signal. For each row the distance error is plotted on a different scale. The columns represent an evaluation without bin width correction of counter values (RAW), with CDT-correction (CDT, see Section 5.3.4) and with bin homogenization (HOM, see Section 5.3.5). Within each row, the same five randomly generated binnings are used – only the type of correction scheme changes from column to column.

Each sub-graph in Fig. 5–42 shows the simulated distance error for input signal phases from 0 to 2π. Each of the five exemplary curves within a sub-graph corresponds to a non-ideal binning that has been randomly generated by the Monte Carlo simulation. Only five and not all of the N random binnings that are generated in course of the simulation are depicted here for clarity.

At high modulation depth (top row in Fig. 5–42), the impact of background light is negligible. All three correction schemes exhibit a relatively small distance error. The exact values are Bosch proprietary. The distance error without correction and the distance error with CDT-correction are of same magnitude despite CDT-correction. The distance error with CDT-correction can be attributed to non-equidistant bin positions and different filter characteristics of the individual sampling windows of unequal width. The simulation clearly shows that, at high modulation depth, bin homogenization is superior to CDT-correction with a significantly reduced distance error.

At medium modulation depth (middle row in Fig. 5–42), the impact of background light considerably disturbs any distance measurement. It should be noted that the distance error is plotted on a different scale on the vertical axis. Without correction a strong distance error can be observed for the exemplary random binnings that renders distance measurements with millimeter accuracy impossible. This error contribution can be attributed to the error due to background light detected with bins of unequal width (see Fig. 5–32 on page 169). At medium modulation depth, the simulated distance errors with CDT-correction and bin homogenization are on par. Regarding the distribution of the distance error, the distance error with bin homogenization shows a modulation for input signal phases from 0 to 2π. A possible explanation is crosstalk from the laser modulator onto the bins as described in Section 5.3.6. Hence, with crosstalk the bin homogenization does not distribute the background light perfectly homogeneously over all counters. This gives a residual background light or modulation-depth-dependent distance error with bin homogenization.

At low modulation depth (bottom row in Fig. 5–42), a reasonable distance measurement without bin width correction is impossible. The distance error is again plotted on a different scale on the vertical axis. Without correction, an error phasor $\vec{E}$ from the non-ideal binning can be larger than a signal phasor $\vec{S}$ (see Fig. 5–2 on page 129). This is the case for the curve reaching diagonal from the top left corner to the bottom right in the graph without correction. At low modulation depth, bin homogenization again suffers even more from crosstalk. CDT-correction, however, nicely equalizes the background light for a measurement at low modulation depth. The prerequisite for successful CDT-correction is that the bin widths have been measured with high accuracy in a calibration measurement. Under these conditions CDT-correction achieves a superior performance than bin homogenization.

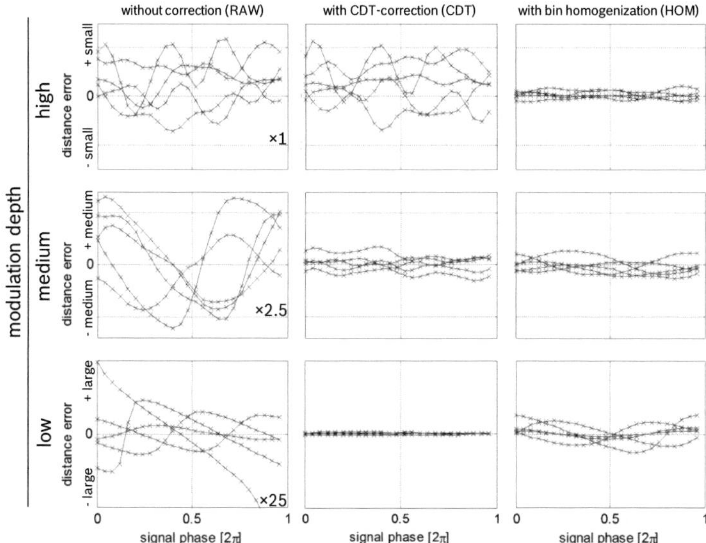

Fig. 5–42: Monte Carlo simulation of binning. Sample of 5 randomly generated binnings. Each sub-graph depicts a simulated distance error over a sweep of the received signal phase from 0 to 2π. The rows correspond to high, medium and low modulation depth of the received signal. For each row the distance error has a different distance error scale, which is then the same for all columns of one row. The columns represent an evaluation without bin width correction (raw counter values, RAW), with CDT-correction (CDT) and with bin homogenization (HOM).

Fig. 5–43 shows the relation between simulated distance error and modulation depth of the received signal for a distance measurement with CDT-correction and bin homogenization. A simulation without correction is not depicted, as it gives too large a distance error at medium distance already and completely fails to measure the correct distance at low modulation depth.

Simulations with CDT-correction result in a rather flat curve over modulation depth. Only towards very low modulation depths, the accuracy with which the bin widths have been determined in a calibration measurement comes into play and increases the simulated distance error. Apart from that, CDT-correction almost perfectly corrects for background light, such that background light only causes a DC offset in the counter values. A DC offset does not impact the phase calculation.

While bin homogenization has troubles at low modulation depths, in particular because of crosstalk as described above, bin homogenization outperforms CDT-correction at high modulation depths. A high modulation depth typically occurs at short distances where a strong signal is received. The impact of crosstalk is highlighted in Fig. 5–43, where the dotted line represents the result of a further run of the Monte Carlo simulation with bin homogenization. All simulation parameters such as disturbances on bin widths and positions are the same for both bin homogenization curves, only the impact of the laser modulator onto the binning has been reduced to 75 % of its previous value. In consequence, the simulated distance error scales by about the same amount.

For the target application of a laser rangefinder a precise distance measurement is most important at short distances, e.g., millimeter precision is required when measuring a marble kitchen table top of 2-3 meters length. This is where bin homogenization shows its strength. In construction, millimeter precision is not that critical for a craftsman. When measuring a 20m distance for a garden fence, even a few cm distance error can then be tolerated. Hence we propose an adaptive system that uses bin homogenization at short distances or at high modulation depth and switching to CDT-correction at low modulation depth.

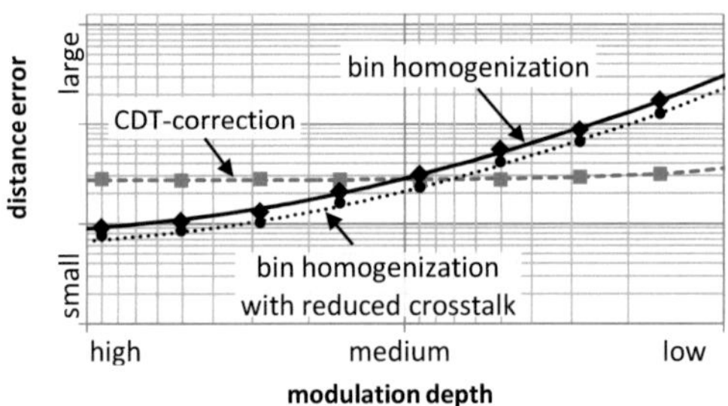

Fig. 5–43: Simulated distance error vs. modulation depth of received signal. Each symbol (square, diamond, disk) corresponds to an operating point of the Monte Carlo simulation. A quadratic fitting curve approximates the simulated operating points.

5.4.2 System Performance Results

The following section presents the distance measurement error for a SPAD-based laser rangefinder. An overview of the performance estimation was shown in the diagram in Fig. 5–1 on page 128. The individual steps have been discussed in the previous sections.

The cumulative error includes influences from the optical system, photon noise, crosstalk, timing drift and the non-ideal binning. The results presented herein are given for the worst-case pixel (see Fig. 5–12 on page 143). For a worst-case pixel, the peak irradiance of the received signal light falls in-between photosensitive (active) detector areas and a minimum amount of light on the photosensitive detector area. Of course it is possible to adjust the receiver optics carefully in a laboratory setup and perfectly focus the laser spot onto the SPAD active area for a specific target distance. Such a setup corresponds to the best-case pixel. For a low-cost handheld laser rangefinder to be operated in rough environment of a construction site, however, this alignment cannot be maintained for the entire lifecycle with anticipated drops on concrete floor, with a plastic support for the optics and temperature variations. Therefore, the following data is presented for the worst-cast pixel as a lower boundary of the expected performance.

Fig. 5–44 presents the simulated cumulative distance error for a target distance of up to 40 m on a linear scale. CDT-correction (CDT) and bin homogenization (HOM) are applied to compensate for non-idealities of the binning. The overall curve shape of the shown in Fig. 5–44 is very similar to the graph shown before in Fig. 5–43. The modulation depth decreases with increasing target distance. A graph of modulation depth versus distance can be found in Fig. 5–22. Despite correction, the overall curve shape is dominated by error contributions from the non-ideal binning. At short distances, where the modulation depth is high, bin homogenization is superior to CDT-correction as shown in the previous section (see Fig. 5–43). At a target distance of about 12.5 m, bin homogenization and CDT-correction are on par. Towards longer target distances, i.e., with decreasing modulation depth, the benefits of CDT-correction become apparent.

The red curve in Fig. 5–44 denotes the target specification defined in Table 1.1 on page 6 (range dependent accuracy, line 3, column 3). A tolerable distance error of 2 mm is specified at short target distance. The error increases to 4 mm at 20 m target distance. At short distances the simulation gives 1.6 mm distance error with bin homogenization and 3.0 mm distance error with CDT-correction. At 20 m target distance the simulation gives 6.7 mm distance error with bin homogenization and 4.2 mm distance error with CDT-correction. The target specification can be met at short to medium target distances up to about 12 m. Bin homogenization is the key for reaching this goal. Even above 12 m target distance, the simulation values with CDT-correction almost reach the target specification. At 20 m target distance and 10,000 lx background light, the simulated laser rangefinder fails to reach the target specification by only 0.2 mm.

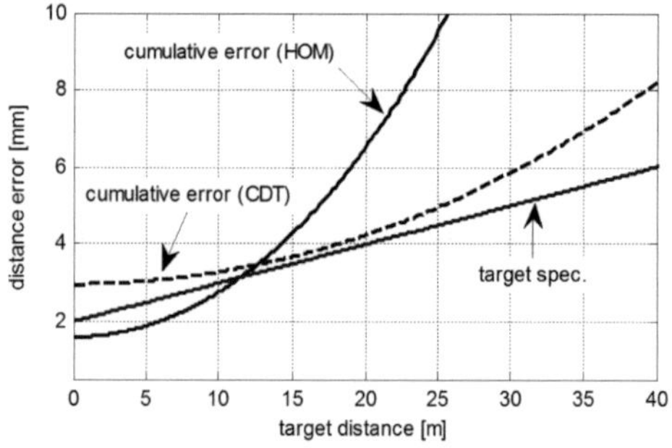

Fig. 5–44: Estimated distance error vs. target distance for a SPAD-based laser rangefinder. CDT-correction (CDT, dashed line) and bin homogenization (HOM, solid line) are applied to compensate for non-idealities of the binning. The red line indicates the target specification (Table 1.1). 1 GHz modulation frequency, 10,000 lx background light.

Fig. 5–45 presents the simulated cumulative distance error for different background light conditions on a double-logarithmic scale. Recall that 100,000 lx correspond to direct sunlight whereas 1,000 lx is the light level in a TV-studio (see Section 5.2.1). At 10,000 lx the simulated SPAD-based laser rangefinder fails to reach the performance target by very little. At 1,000 lx background light however, CDT-correction reaches a distance error of 3.3 mm and meets the specified distance error. With bin homogenization the performance is further improved to a cumulative distance error of 2.2 mm.

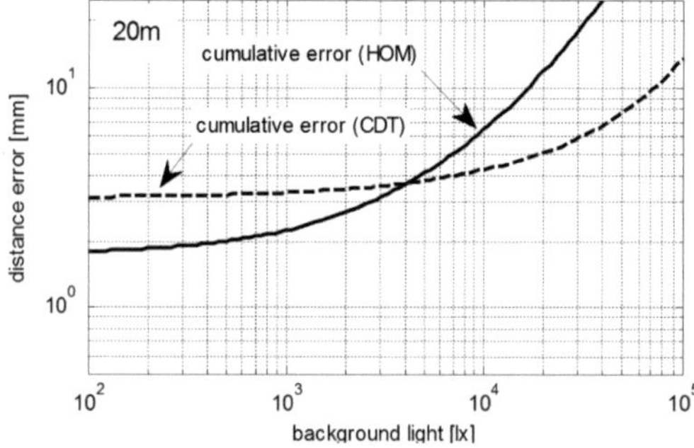

Fig. 5–45: Estimated distance error vs. background light for a SPAD-based laser rangefinder. CDT-correction (CDT, dashed line) and bin homogenization (HOM, solid line) are applied to compensate for non-idealities of the binning. 1 GHz modulation frequency, 20 m target distance.

Because of the susceptibility to background light, a SPAD-based laser rangefinder is not well-suited for outdoor applications with operation in direct sunlight. The system should be considered an indoor device at least at medium to long distances. At short distances, for example measuring the width of a window sill, the system is capable of tolerating a certain amount of background light.

Lastly, Fig. 5–46 present the simulated distance error versus modulation frequency at 20 m target distance and 1000 lx background light. As described previously, photon noise (see Section 5.3.2.4) and optical and electrical crosstalk (see Sections 5.2.4.3 and 5.3.6) cause a phase error. A given phase error causes a frequency dependent distance error that scales with the inverse of the frequency (1/f) (Eq. (2.10)). This 1/f drop is the dominant influence at low frequencies. At a modulation frequency of 30 MHz, as used in [39], our simulation gives a distance error of 89 mm. Towards higher modulation frequencies of some hundred MHz, the frequency dependent distance error is gradually superseded by distance errors due to delay drift in the binning between target and reference path as well as estimated delay drifts from the SPAD array to the binning (see Section 5.3.7).

At a modulation frequency of about 800 MHz, a cumulative distance error of 2.2 mm is reached with bin homogenization. Further increasing the frequency does not significantly reduce the distance error any further, because the distance error is now dominated by delay drifts (Section 5.3.7). On the contrary, a further increase in frequency can be counter-productive, because the filter characteristic of the system will further attenuate the modulation amplitude towards higher frequencies (see Section 5.3.2.2, Fig. 5–19). A lower modulation amplitude in turn has negative impact on the distance error (see Fig. 5–43). The optimum frequency of operation, however, depends heavily on the design and implementation of the actual system.

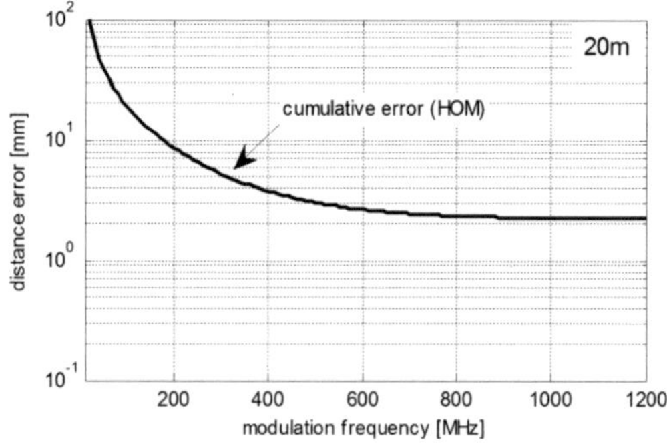

Fig. 5–46: Estimated distance error vs. modulation frequency for a SPAD-based laser rangefinder. Bin homogenization (HOM) is applied to compensate for non-idealities of the binning. 20 m target distance, 1,000 lx background light.

5.5 Conclusion

Based on the system performance estimation, a SPAD-based laser rangefinder is an attractive candidate for a low-cost system.

The results of the optical analysis show that the expected light levels – and thus count rates - for a laser rangefinder lie well within the operating conditions for the single-photon avalanche diodes characterized in Chapter 4. Count rates up to 11.5 MHz are expected. A sufficient modulation depth of the received signal light on top of background light is reached at short to medium target distances. Burst modulation (see Section 5.2.3.2) can further improve the signal-to-noise ratio. An eye safe average power of 1 mW of the transmitter is assumed, with a low-cost laser diode as an appropriate light source.

The system performance depends on how well the image of the laser spot is focused on the SPAD active (photosensitive) area. As alignment of the optical system cannot be maintained with µm-precision in the harsh environment of a construction site, a worst-case scenario is assumed. In this case, the photosensitive area only receives light from the periphery of the laser spot and the peak irradiance falls onto the non-photosensitive area of the SPAD pixel required for electronic circuitry (see Fig. 5–12). Alternatively the image of the laser spot can be widened by defocussing.

Four major types of distance error determine the performance estimation: photon noise, crosstalk, binning non-idealities and temporal drift. Photon noise in single-photon synchronous detection (SPSD) leads to a phase error. Therefore, the modulation frequency should be as high as possible so that the fixed phase error corresponds to a small distance error. Optical crosstalk and electrical crosstalk on the photon detection efficiency (PDE) of the SPAD also cause a phase error. Also in this case a high modulation frequency is desired. Nonetheless, the filter characteristic of the SPAD jitter reduces the signal modulation towards high frequencies and thus sets an upper frequency limit to the optimum operating range. For the simulated system, the sweet spot lies in the range of 800 MHz to 1 GHz. Of course, the optimum frequency range depends on the specific type of SPAD and binning architecture, and can be different in other semiconductor technologies.

A binning architecture based on a ring-voltage-controlled-oscillator (ring-VCO) has been developed to enable operation at modulation frequencies up to 1 GHz. Recalling that the employed semiconductor process is an imaging technology that has not been optimized for RF performance. The binning derived from this ring-VCO suffers from non-idealities: unequal bin widths and bin positions. A Monte Carlo simulation shows that distance measurement is not possible without correction. Two bin width correction schemes are presented:

- CDT-correction
  In a first step the non-ideal bin width is determined in a calibration measurement prior to the actual distance measurement. In a second step the counter values recorded during distance measurement are divided by the bin widths determined in the calibration measurement.

- Bin homogenization
  The measurement time is split into sub-measurements of equal duration. The number of sub-measurements corresponds to the number of bins. Furthermore, the fixed relation between bin and counter is broken up. For each sub-measurement, a counter is incremented by a different bin. Thus, each bin increments each counter for the duration of one sub-measurement. Thereby, despite non-ideal bin widths, the background light is homogeneously distributed over all counters. Nonetheless, bin homogenization maintains synchronization between the modulated signal and the counters. A calibration measurement is not required at all, however, the conditions between sub-measurements must be stable.

The selection of the appropriate correction scheme depends on the operating conditions. At strong crosstalk and low modulation depth (i.e. strong background light and/or distant target) CDT-correction shows better performance, whereas bin homogenization shoes superior performance in the near range. The following table briefly summarizes the strengths and weaknesses:

CDT-Correction	*Bin Homogenization*
⊖ Calibration measurement mandatory, however calibration can mitigate crosstalk ⊖ Higher distance error at short range (0 to 12.5 m target distance) ⊕ Lower distance error at low signal modulation depth	⊕ No calibration measurement required ⊕ Lower distance error at short range / high modulation depth ⊖ Susceptible to crosstalk, therefore higher distance error at low signal modulation depth

Table 5.2: Comparison of CDT-correction and bin homogenization

The temporal drift of the transmitter requires compensation such as a device internal reference measurement. The temporal drift is predominantly temperature dependent. An additional reference SPAD can easily be integrated within the same ASIC at low cost. The absolute temporal drift of target and reference path does not matter, because only the difference is evaluated. Thus, a substantially identical, symmetrical layout of the circuitry of the electronic target and reference path on the ASIC is crucial for identical drift behavior and minimum mismatch between both paths.

Considering all error contributions, the results of the performance estimation prove that it is possible to meet the target specifications for a low-cost laser rangefinder (outlined in Table 1.1) up to 12m target distance at 10,000 lx background light. The low distance error at a short target distance is achieved using the proposed bin homogenization. At longer target distances, the specified distance error can only be reached at lower background light levels. Nevertheless, the estimated 4.2 mm distance error at 20 m target distance using CDT-correction misses the target specification by only 0.2 mm.

In conclusion, a SPAD-based system is expected to work as a platform for an low-cost laser rangefinder. The estimated system performance of this low-cost system, however, is inferior compared to state-of-the-art high-priced APD-based laser rangefinder. Furthermore, application scenarios will be limited to some 20m range and indoor measurements. High precision measurements under outdoor conditions will remain the domain of the APD-based laser rangefinder.

6 Rangefinder-ASIC Design and Characterization

This chapter presents the experimental results of two ASICs for a SPAD-based laser rangefinder. The target of this chapter is to demonstrate the feasibility of a millimeter-precision laser rangefinder using a low-cost photon counter (SPAD).

The requirements and specifications for the ASICs have been developed based on the experimental results of SPAD measurements (Chapters 4) and considerations regarding the laser rangefinder system (Chapter 5). Both ASICs are designed by Bosch Automotive Electronics according to these specifications. SPADs are provided by STMicroelectronics. The ASICs have been fabricated by STMicroelectronics.

A first version of a laser rangefinder ASIC (CF340 MPW2009, ASIC 1) has been implemented on a first multi-project wafer (MPW2009). Compared with the state of the art, this is the first ASIC for distance measurement which includes SPADs, timing-circuitry (binning) and laser driver on a single ASIC. Furthermore, the ASIC is implemented in a well-established fabrication process for mass-production. Experimental results for the CF340 MPW2009 are presented in Section 6.1.

Based on the experimental results with the first ASIC, a follow-up design has been implemented on a second multi-project wafer (MPW2011). This second laser rangefinder ASIC is referred to as CF340 MPW2011 (ASIC 2). The second ASIC comprises several improvements, in particular SPADs with improved fill factor, a larger SPAD array, a phase-locked loop (PLL) with better matching delay stages and reduced pulse gaps of the binning, and better shielding of the binning from crosstalk from the laser driver. The new ASIC also includes bin-homogenization for background light suppression without prior calibration as proposed in Section 5.3.5. Experimental results for the CF340 MPW2011 are presented in Section 6.2.

A special focus is set on the binning performance which is analyzed in detail. Although the performance estimation in the previous chapter is based on timing uncertainties estimated from individual substructures of the ASIC, it is not possible to analyze these substructures individually in an experiment. Often only a group of functional blocks is available. It is in the nature of integrated circuits that access from the outside world is limited to a few pins for electrical testing and a few output quantities, i.e., counter readings, from which the parameters of interest have to be derived.

6.1 Characterization ASIC 1 (CF340 MPW2009)

6.1.1 Block Diagram and Evaluation Board

A schematic block diagram of the CF340 MPW2009 (ASIC 1) is shown in Fig. 6–1. The chip comprises a phase-locked loop (PLL) for RF generation from an external 10 MHz reference frequency; an on-chip laser driver for direct modulation of the laser diode; a 3×3 array of SPADs for detection, and the binning for sampling the received signal. SPAD pulses can optionally pass through the binning to a test output pin. Pulses from the individual bins are sorted into eight 32 bit counters and read out via an i^2c interface. The PLL covers a frequency range from 400 MHz to 1200 MHz. Control logic is not depicted.

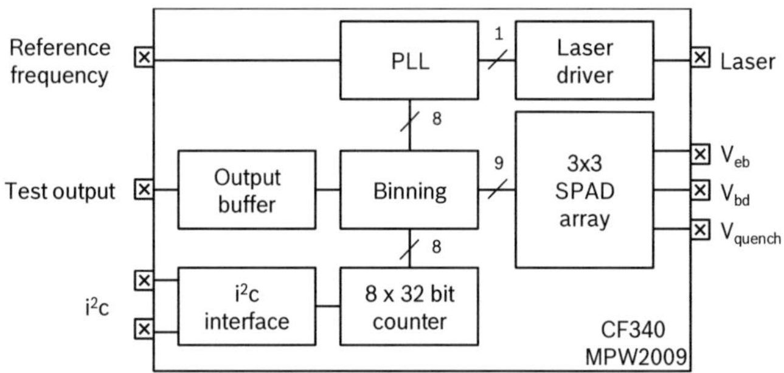

Fig. 6–1: Block diagram of CF340 MPW2009

A photomicrograph of the chip is shown in Fig. 6–2.

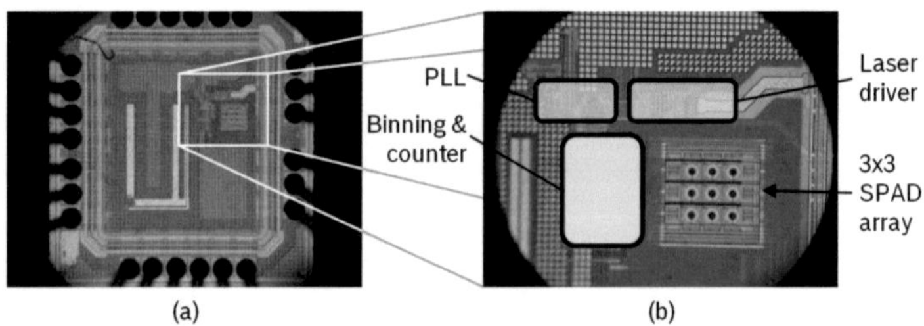

Fig. 6–2: Photomicrograph of CF340 MPW2009 test chip (a) overview and (b) main building blocks of laser rangefinder at 125x magnification. Details of the binning architecture are Bosch proprietary.

An evaluation board has been designed by the author to test the CF340 MPW2009 in a laser rangefinder setting (Fig. 6–3). Besides the ASIC, the evaluation board includes a low-cost microcontroller (µC) for communication with the ASIC and a host computer; an off-the-shelf laser diode emitting at 635 nm wavelength; a 10 MHz quartz to generate the reference frequency, and ancillary electronics such as voltage regulators, filter elements and bias generation for the laser diode that are not specifically labeled in Fig. 6–3. An LED for background light generation (not shown) is also controlled by the evaluation board. SPAD operating voltages are supplied by external source-measure-units (Keithley 2602B). High voltages (up to 15 V) for the SPAD supply are currently not generated on chip, however, concepts for on-chip HV-generation exist and are about to be implemented alongside CMOS SPADs [130].

The entire ranging system can be easily scaled down to few square centimeters of printed-circuit-board (PCB) area as required for a handheld device. Processing of the counter values that represent SPAD counts is performed either with the low-cost µC or on an external computer running Matlab.

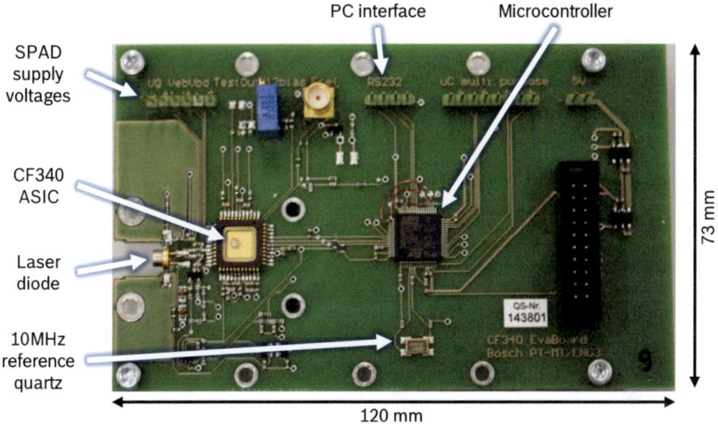

Fig. 6–3: Photograph of CF340 MPW2009 evaluation board

6.1.2 Signal Light

An off-the-shelf red laser diode (635nm) serves as the signal light source on the evaluation board. Such laser diodes are widely used in low-cost commodity products such as CD-players. The beam is collimated with a plastic lens collimator from a laser rangefinder already available on the market. An onboard control circuit ensures 1 mW mean output power for laser class II eye-safe operation. The receiving optics in our laboratory setup is a 25 mm diameter spherical lens with 30 mm focal length (Fig. 6–13 on page 211).

6.1.2.1 Modulation and Spectrum

Two major aspects for the performance are the modulation depth of the signal and its spectrum. Fig. 6–4 (a) depicts the optical output power of the laser diode as a function of time. The optical signal is measured with a high-speed photodiode (4.5 GHz bandwidth). An oscilloscope with an analog bandwidth of 5 GHz samples the output at a sampling rate of 20 GHz. The corresponding spectrum is calculated from the samples and is shown in Fig. 6–4 (b).

For a sinusoidal intensity modulation the laser should be operated sufficiently above lasing threshold. The implemented laser driver taken from an already existing laser rangefinder, however, goes below lasing threshold during modulation and thereby introduces distortions. The optical signal is rather a pulse train with a strong presence of higher harmonics. Nonetheless a pulse train can be advantageous since higher modulation depths can be reached (as shown in Appendix D). The modulation depth on the fundamental modulation frequency in Fig. 6–4 is $m_{\text{signal}} = 146\%$.

A more detailed analysis at different modulation frequencies showed that the pulse shape does not change significantly with the modulation frequency. This can be attributed to the optical pulse being the first oscillation from relaxation oscillations of the laser.

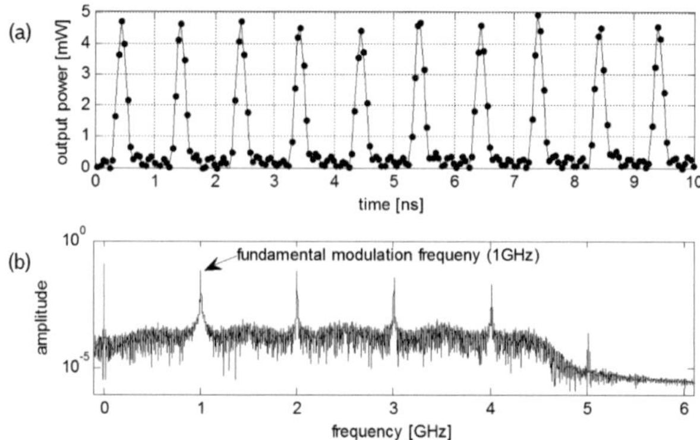

Fig. 6–4: Laser driver modulation (a) optical power in time domain (b) calculated spectrum of optical signal having a fundamental modulation frequency of 1 GHz. Captured at 20GSa/s with a high-speed photodiode (4.5 GHz bandwidth).

6.1.2.2 Speckles

When going to small detector areas, speckles (Section 5.2.3.3) become relevant. Any change of the target surface will create a different interference pattern in the detector plane. Since speckles also depend on the receiver optics, any change in the optical system also causes different speckle patterns.

In this experiment, a collimated laser beam is sent out onto a target at a fixed distance. The target is a white sheet of paper, thus a rough surface on wavelength scale. A stepper motor moves the target between measurements orthogonal to the optical axis. Fig. 6–5 (a) shows the counts received by one SPAD for 200 repeated measurements of same measurement time. Even though the laser power is fixed, the measured counts change strongly for different target positions. The relative photon noise from Poisson distributed counts at about $\overline{N} = 5 \times 10^6$ counts is $1/\sqrt{\overline{N}} = 1/\sqrt{5 \times 10^6} = 0.04\%$ (Eq. (3.13)) and therefore negligible.

The same measurement is repeated with the target constantly rotating about the optical axis. Thereby, we average over a plurality of speckle situations during each measurement. With the rotating target we obtain a substantially constant count value, as shown in Fig. 6–5 (b). The rotating target is used in all further experiments.

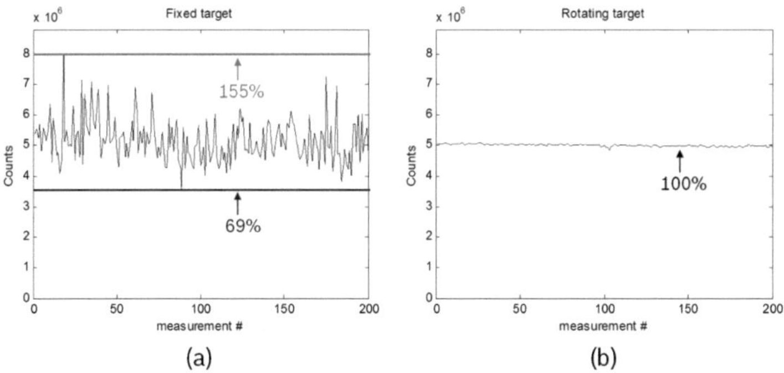

Fig. 6–5: Measured counts for different speckle situations. Laser spot on target at fixed distance measured with 8 µm diameter disk-shaped SPAD, receiving optics with 8 mm aperture and lens with 30 mm focal length (a) repeated measurements for laser spot on different points on the target with speckles (b) repeated measurements for rotating target without speckle effects.

A measurement of different locations on the target actually represents a measurement with a handheld laser rangefinder in practice. Without a tripod a user trembles slightly when pointing at a target at several meters distance – and thus averages over different speckle patterns. In this case the measurement time of 1-4 s is long compared to the time scale on which the speckle pattern changes (user typically trembles at a frequency of some Hz).

6.1.3 Binning

The binning is analyzed in this section. The binning is the core element for sampling the received signal in synchronism with the laser modulation.

6.1.3.1 Bin Width

Bin width is the most important measure in terms of background light tolerance. As long as the background light (BL) is homogeneously distributed over all bins, BL will not cause a systematic distance measurement error. In that case BL only constitutes a DC offset. The absolute bin width b_k for each bin k is not needed for BL compensation but rather the *relative* bin width $w_{\mathrm{rel},k}$ as defined in Section 5.3.4, Eq. (5.29).

Fig. 6–6 exemplarily shows the relative bin widths measured for one device. The device is operated at a modulation frequency of 1 GHz. The chip is illuminated with constant background light, i.e., no modulation signal is present. It should be noted that a 100 Hz flicker from ceiling lighting does not disturb the measurement as this is slow with respect to the modulation frequency. Bin widths are determined from a large number of photons N (millions of photons per bin). The relative accuracy (see Eq. (3.13)) with which the bin width is determined scales with $1/\sqrt{N}$, thus for a large number N the error bar is negligible and omitted in the graph. It results from Fig. 6–6 that the bins do not have the same width.

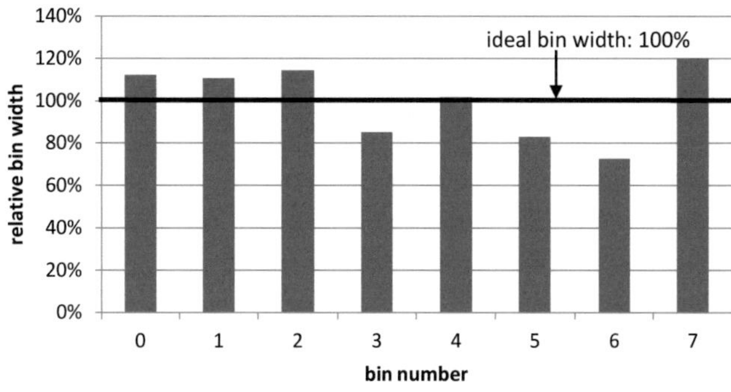

Fig. 6–6: Bin width. Sample 4 at 1 GHz frequency

It has been shown in the previous chapter in Section 5.4 that BL compensation is mandatory to reach the performance target. An advantageous scenario for mass production would be that all chips show exactly the same binning non-idealities. Ideally, a one-time calibration is sufficient for all chips at any modulation frequency and any temperature over the entire device lifetime. The worst-case scenario is a binning that must be calibrated before each measurement or even before sub-measurements at different modulation frequencies.

Fig. 6–7 exemplarily shows the *absolute* bin widths for multiple devices operating at a modulation frequency of 1 GHz. The bin width is different for each sample. This time the counts x_k from each bin k are compared with counts x_external that bypass the binning and are recorded at the test output with a sampling oscilloscope or external counter for an identical measurement time. With the period $T = 1/f_m$, this gives an absolute bin width

$$ b_k = \frac{x_k}{x_\text{external}} T . \tag{6.1} $$

At a modulation frequency of 1 GHz, the ideal bin for eight bins per period spans 1/8 of the modulation period, i.e., 125 ps. As can be seen in Fig. 6–7 the measured bins are on average smaller than the ideal bin width. This effect can be attributed to the semiconductor fabrication process used for SPADs, which is optimized for imaging but not for routing RF signals over the chip. High parasitic capacitances and a high line resistance decrease the slope and pulse amplitude of the signals for opening and closing the bins and thus cause a pulse gap between bins (see Fig. 5–30). At lower modulation frequencies, the pulse gap diminishes as can be seen in Fig. 6–8. It should be noted that the reduced bin width must be taken into account when estimating system performance and noise error since fewer SPAD pulses are detected. A longer measurement time is required for compensation.

Even the day-to-day variation of environmental conditions in the air-conditioned laboratory yields slightly different bin widths. The measured counts for repeated measurements with the same chip at different days exceeds the expected statistical count variation. Nevertheless, Fig. 6–7 shows a certain systematic behavior. Bin 6 for example is always smaller than an average bin. This can be attributed to a non-ideal design of the ASIC with unbalanced loads or parasitics in the VCO elements or in the routing towards the binning (Section 5.3.3.4, page 166).

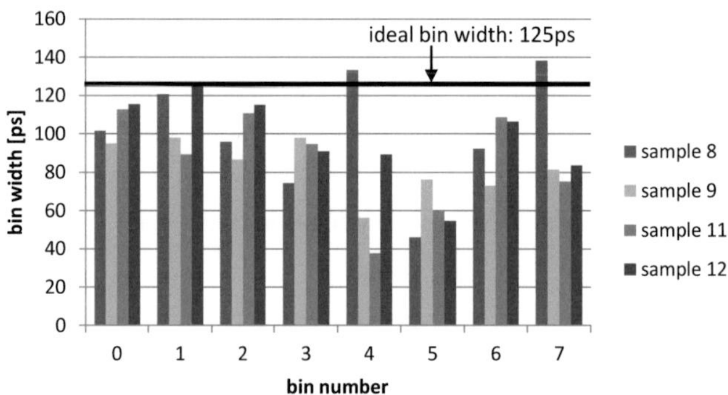

Fig. 6–7: Absolute bin widths. Multiple CF340 MPW2009 samples.
Measurement at 800 MHz modulation frequency.

Bin widths are also analyzed over the entire frequency range of the PLL in Fig. 6–8. Absolute bin widths are measured and normalized to 100% for a bin of ideal bin width of 1/8 of the modulation period at each frequency. The absolute normalized bin width $b_{k,\text{normalized}}$ for 8 bins is given by

$$b_{k,\text{normalized}} = \frac{x_k}{\frac{1}{8} x_{\text{external}}} \; .$$

(6.2)

As expected, the influence of timing differences between opening and closing edges of different bins reduces towards lower modulation frequencies. Unfortunately, a separate bin width compensation must be applied for each modulation frequency in case of using a code density test (CDT) since the bin width changes. It should be noted that bin 7 does not open at all above 1.1 GHz, which can cause a problem (division by zero) in simplistic correction algorithms.

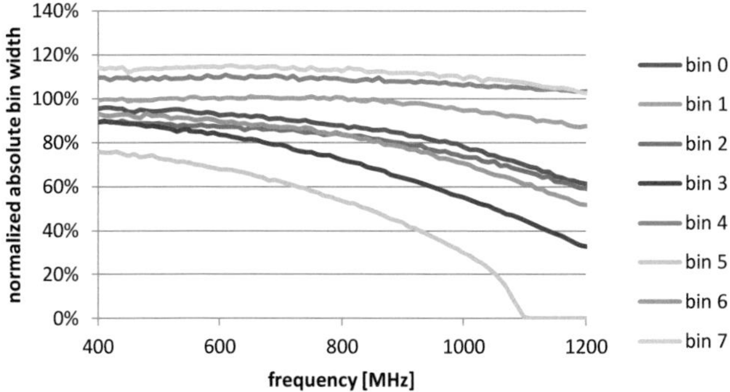

Fig. 6–8: Bin width vs. frequency. The bin width is normalized so that 100% represent 1/8 of the modulation period. Sample 8.

6.1.3.2 Bin Position

Ideal bin centers for N bins are positioned equidistantly at intervals $2\pi/N$. The control signals for the bins are derived from the VCO-stages of the PLL (see Fig. 5–29). A mismatch of the delay elements not only influences the bin width but also the bin position in time as described in Section 2.3.3.1 with reference to Fig. 2–14. The Monte Carlo simulation in the performance estimation in Section 5.4 also accounts for disturbed bin positions.

Fig. 6–9 exemplarily shows the deviations of measured bin positions with respect to ideal bin positions for one sample. The binning operates at a modulation frequency of 1 GHz. Bin 2 deviates by 14 ps from the ideal bin position. The bin is thus shifted by 1.4% of the RF period. Small bins such as bin 4, 5, 6 in Fig. 6–7 indicate that the delay time of the respective VCO stage is too short. The bin position will in turn start ahead of time such as bin 5, 6 in Fig. 6–9.

The measurement of bin positions is not feasible in a handheld laser rangefinder but requires a precise laboratory setup. Distance measurements presented in this thesis are only corrected for bin width, and no correction for bin position was employed.

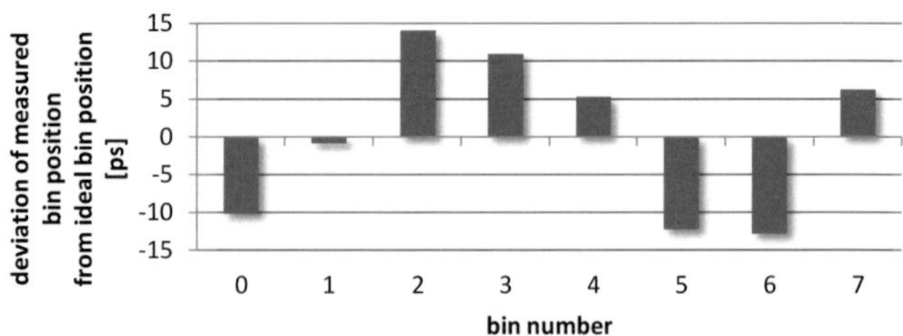

Fig. 6–9: Bin positions. Deviation of measured bin position from the ideal bin position for equidistant sampling every $2\pi/N$ for N=8 bins.

6.1.3.3 Temperature Drift

As a laser rangefinder is used on a construction site indoors and outdoors, the binning must function over a temperature range from -10°C to 50°C. In this experiment, the chip is placed over a heating plate during a temperature cycle. The bin width for one exemplary bin during the temperature cycle is shown in Fig. 6–10. The relative bin width changes by 2.97 %. As a consequence, it is not sufficient to perform a one-time calibration of the bin width. A separate bin width correction is mandatory at different temperatures.

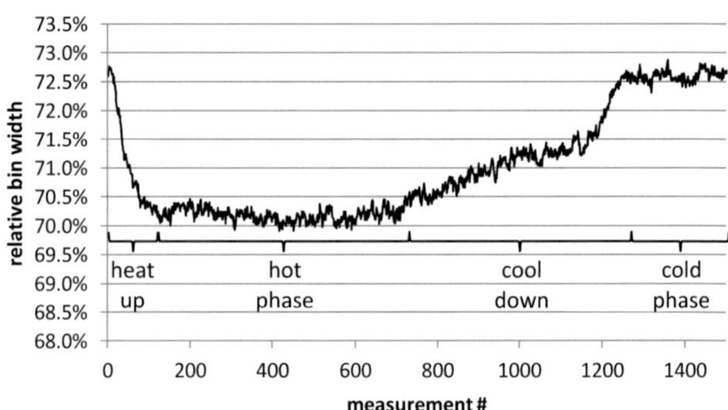

Fig. 6–10: Bin width during a temperature cycle.
Sample 4, bin 6 at 1 GHz frequency.

6.1.3.4 Crosstalk

Integrated systems are especially prone to mutual influences of system elements. While on-chip signals operate at minimum currents, the laser modulator drives the external laser diode with a current of several milliamperes. This causes strong crosstalk (a) on the opening and closing edges of the binning, thus influencing the bin width and (b) on the supply voltages of the SPAD, which in turn changes the photon detection efficiency (PDE) (Section 5.3.6).

Fig. 6–11 illustrates the combined effect of crosstalk from the laser modulator on binning and PDE. The SPADs are illuminated with constant background light and the counts recorded for a measurement period of time. A first measurement with the laser modulator turned off defines the baseline for a code density test (CDT, see Section 5.3.4) for correcting the recorded counter values. The counter values measured with increasing laser modulator currents (low, medium, high) are divided by the counter values obtained with the laser modulator turned off. A strong modulation over the eight bins is observed.

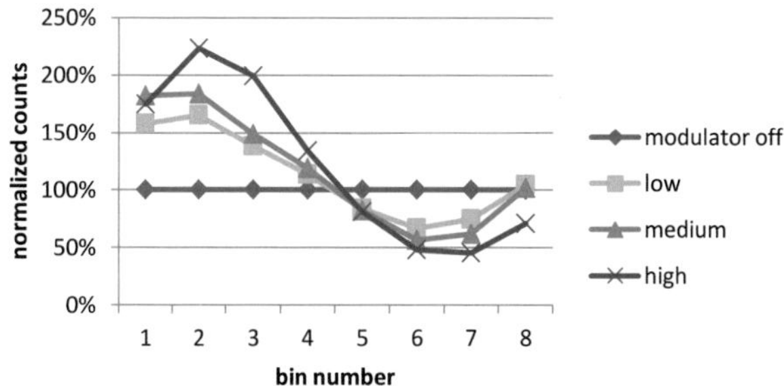

Fig. 6–11: Crosstalk measurement. A code density test (CDT) based on the measurement with the on-chip laser modulator turned off is applied to the subsequent measurements with increasing laser modulator currents (low, medium, high). 800 MHz modulation frequency.

The first ASIC (CF340 MPW2009) has not been optimized with respect to crosstalk yet. One component that needs a redesign is the chip package / bonding scheme. For example, the lengthy laser driver output bond wire connect to the ASIC in parallel directly adjacent to the SPAD supply voltage lines. This opens a path for coupling that can be avoided by design. Furthermore, on-chip filtering and external filter components on the evaluation board leave room for improvement.

6.1.4 Distance Measurement

The measurement principle for time-of-flight distance measurements has been outlined in Chapter 2.2. The laser rangefinder system, including optical components, is described in Chapter 5.

The strong crosstalk (see above, Section 6.1.3.4) renders a phase measurement with a single chip useless. Even at low background light levels unsatisfactory results are expected. However, two separate but identical chips can be used instead of a single chip. The first chip serves as the transmitter (Tx), which modulates the laser, and the second chip serves as the receiver (Rx) with SPAD array and binning for detection and sampling of the received signal. Both chips can be synchronized with an external reference frequency generator.

6.1.4.1 Measurement Setup

Basically all required electronic components for a laser rangefinder are integrated in the CF340 MPW2009 ASIC. This favorably reduces the laboratory setup to a minimum. Fig. 6–12 shows a simplified sketch of the distance measurement setup. The first chip is mounted on a Tx Board, the second chip is mounted on an Rx board. Both boards are connected to a reference frequency generator. A white sheet of paper is chosen as the non-cooperative target surface which diffusely backscatters incident light. To avoid speckle effects (see Section 6.1.2.2), the target plate is mounted on a motor that rotates the target about to the optical axis. The target fixture in turn is mounted on a linear stage with 2 m length which is placed at 1.2 m distance offset from the boards. The target distance can thus be varied from 1.2 m to 3.2 m in steps of 12.5 µm with ±20 µm positioning repeatability. A spherical lens without near-range element (Section 5.2.4.2, text after Eq. (5.13)) is used as the receiver optics.

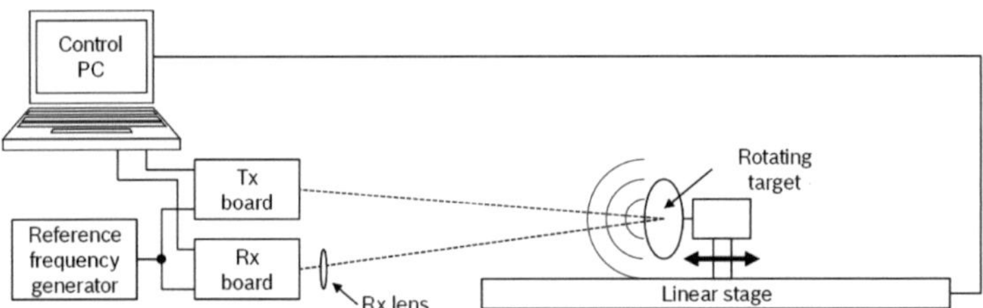

Fig. 6–12: Sketch of distance measurement setup

Fig. 6–13 shows a photograph of the 'laser rangefinder' part of the distance measurement setup.

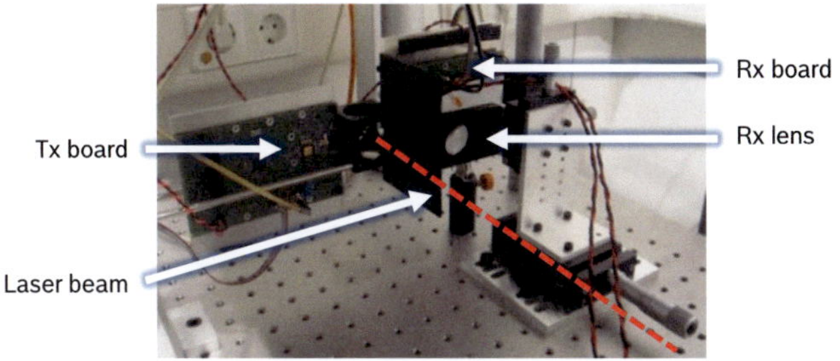

Fig. 6–13: Distance measurement setup. Photograph of Tx and Rx board with optics.

The measurement range in our laboratory is limited to a maximum target distance 3.2 m. While Fig. 6–12 only shows a simplified sketch, the actual laboratory setup offers more degrees of freedom. By means of additional attenuators (optical attenuation filters) in the sending and receiving optical paths and by controlled background light sources, we can simulate operating conditions that correspond to measuring a much longer range by choosing, e.g., low modulation depth and high background light level. Different scenarios are presented for the second ASIC (CF340 MPW2011) in Section 6.2.3.

6.1.4.2 Measurement Result

A distance measurement is performed at a modulation frequency of 1 GHz. At each distance the train of SPAD pulses is sampled by the binning in synchronism with the laser modulation and increases the corresponding counters. The phase of the received signal is calculated from the Fourier transform of the counter values. Measurements are performed at 1 cm target distance increments with 1 s measurement time at low background light and high modulation depth. At a modulation frequency of 1 GHz, the unambiguity range is 15 cm. Though only measurements at a single frequency are presented for clarity, measurements at multiple closely spaced frequencies extend the unambiguity range (see Section 2.2.3). Other frequencies can easily be programmed by digitally setting a different PLL divider.

Fig. 6–14 depicts the relationship between measured phase and true target distance. The measured phase in one unambiguity range changes from $-\pi$ to $+\pi$. In a next step the phase is unwrapped, i.e., 2π phase jumps are removed. The distance is then calculated from the unwrapped phase using Eq. (2.9). Non-idealities in the measurement system cause a deviation between distance obtained from a measured phase and true target distance. The label 'RAW' denotes the phase of the received signal calculated from the counter values without bin width correction. The label 'CDT' gives the calculated phase after correction of the measured counter values using the bin widths obtained from a prior code density test (CDT).

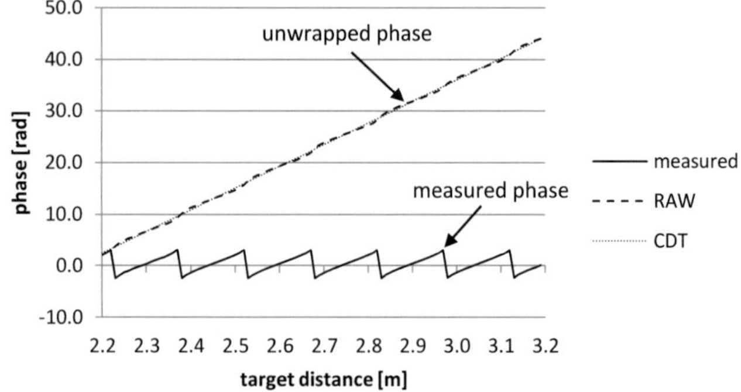

Fig. 6–14: Phase vs. true target distance. 'RAW' denotes the phase at the fundamental modulation frequency calculated from the counter values without bin width correction. 'CDT' gives the result after correction of counter values using bin widths from a code density test (CDT). The unwrapped phase is obtained from the measured phase by eliminating 2π phase jumps.

Deviations between measured target distance and true target distance are plotted in Fig. 6–15 as the distance error. Recall that a distance error of 1 mm corresponds to a phase error of 42 mrad at a modulation frequency of 1 GHz. For the 'RAW' distance measurement without bin width correction, we obtain a distance error of 10 mm. The periodicity of the distance error corresponds to the modulation period. With CDT-correction, the systematic error reduces to a maximum distance error of 0.6 mm. The target of building a millimeter-precision laser rangefinder using a low-cost photon counter is successfully accomplished.

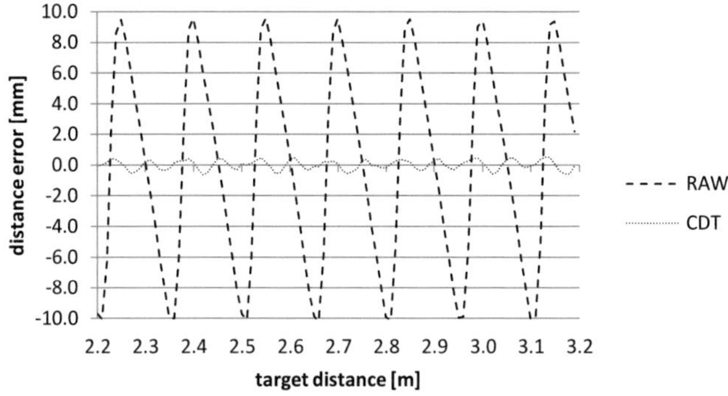

Fig. 6–15: Distance error for distance measurement. Without bin width correction (RAW) and with bin width correction via code density test (CDT).

The achieved accuracy represents a factor of 170 improvement in distance accuracy for single-photon synchronous detection. The authors in [9] report a 110 mm systematic error at 2.4 m target distance under laboratory conditions. As shown in Eq. (2.10), the distance error is inversely proportional to the modulation frequency. The different modulation frequency of 1 GHz in our experiment compared to a modulation frequency of 30 MHz in [9] therefore accounts for an improvement factor of 33.

The code density test (CDT) comes at the price of measuring the bin width at high accuracy prior to the actual distance measurement. In the present experiment, a calibration measurement of 10 s duration with an on-board background light LED to ensure sufficiently high count rates accurately determines the bin width before the actual distance measurement. This drawback is overcome by the bin homogenization scheme implemented in the CF340 MPW2011 (ASIC 2) which presents a powerful method of bin width correction *without the need for prior calibration.*

Noise contributions at different lighting scenarios will be discussed further below for measurements with the CF340 MPW2011 (ASIC 2).

6.2 Characterization ASIC 2 (CF340 MPW2011)

Experiments with the CF340 MPW2009 (ASIC 1) led to an improved follow-up design, CF340 MPW2011 (ASIC 2). This second ASIC comprises several improvements, in particular SPADs with improved fill factor, a larger SPAD array, a phase-locked loop (PLL) with better matching delay stages and reduced pulse gaps of the binning, and better shielding of the binning from crosstalk from the laser driver. The new ASIC also includes bin-homogenization for bin width correction without prior calibration as proposed in Section 5.3.5.

6.2.1 Block Diagram and Evaluation Board

Fig. 6–16 shows a block diagram of the CF340 MPW2011. It comprises a phase-locked loop (PLL) for RF generation from an external 10 MHz reference frequency; an on-chip laser driver; a 6×10 array of SPADs and the binning for sampling the received signal. New SPAD structures (see Section 4.2) are implemented. SPAD supply voltages are provided off-chip. Pulses from individual SPAD can be switched to a test output pin for external measurements. The eight 32 bit counters are read out via an i^2c interface. New additional building blocks are the Tx and Rx phase-shifters, which constitute the hardware implementation of the bin homogenization scheme (see Section 5.3.5 for bin homogenization and Section 5.3.5.3 on page 181 for the implementation).

The phase-shifters are rather switches than delay elements. The Tx phase-shifter subsequently selects different output signals of the VCO stages (see Fig. 5–28) as an input of the laser driver. The output signals of the VCO stages are on average delayed by 1/8 of the modulation period with respect to the previous VCO stage at any modulation frequency. The eight inputs to the Tx phase-shifter in Fig. 6–16 represent the eight different output signals of the VCO stages whereas the one output of the Tx phase-shifter is the selected one with the desired delay.

Similarly the Rx phase-shifter is a logic element that can route the SPAD pulses falling into one bin of the binning towards any selected counter. It should be noted that the binning is hard-wired to the VCO stages of the PLL. Thereby each bin maintains its (possibly non-ideal) width regardless of the state of the Rx phase-shifter. The Rx phase-shifter only changes the association between bin and counter according to the bin homogenization scheme shown in Fig. 5–36 on page 180. This change does not impact the sampling process since the routing from the bin to the counter is not on the time-critical path. A control logic (not shown) synchronously controls Tx and Rx phase-shifters to perform bin homogenization.

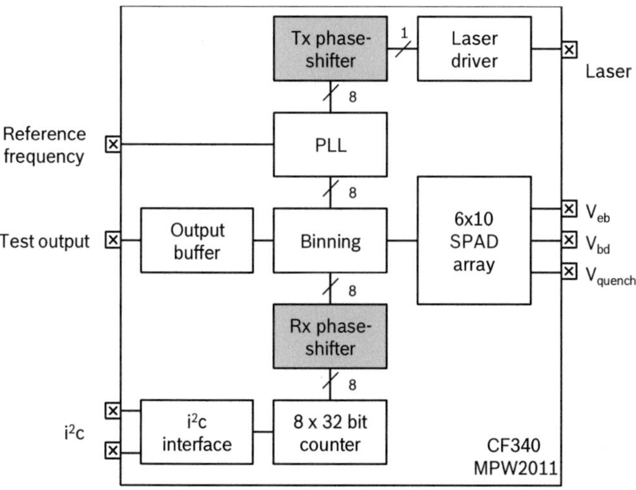

Fig. 6–16: Block diagram of CF340 MPW2011. The shaded blocks are new as compared to the block diagram of CF340 MPW2009 shown in Fig. 6–1.

We designed and fabricated an evaluation board to test the CF340 MPW2011 in a laser-rangefinder setting (Fig. 6–17). The evaluation board comprises a low-cost microcontroller (μC) for communication with the ASICs and a host computer; an off-the-shelf laser diode emitting at 635 nm wavelength; a 10 MHz quartz to generate a stable reference frequency shared among both ASICs and the μC; and ancillary electronics that are not specifically labeled. SPAD operating voltages are supplied by external source-measure-units (Keithley 2602B). The USB interface can provide both communication interface and 5V power supply.

Because of crosstalk between sending and receiving path on chip, two separate evaluation boards were being used in previous measurement setups with the first ASIC, CF340 MPW2009 (see Fig. 6–12). Both boards were synchronized with an external frequency generator. Because the CF340 MPW2011 shares the same pad-ring (i.e., the arrangement of contact pads for contacting the chip with bond wires), bonding scheme and several blocks with the CF 340 MPW2009, a certain amount of crosstalk can be expected for the new setup as well. Despite optimizations in PLL and binning, there are several coupling paths such as the parallel bond wires for laser driver output and SPAD supply lines that also exist in the new setup. Therefore two CF340 MPW2011 ASICs are placed on the evaluation board. Both chips are synchronized with the reference frequency from the on-board quartz. One (Tx) chip can be used for modulating the laser while the other (Rx) chip is used solely for detection with its laser modulator turned off. See Section 6.2.2.3 for crosstalk analysis.

Laser and driving circuitry are similar to the previous evaluation board. The analysis in this section will therefore focus on binning, the bin-homogenization scheme, and distance measurements.

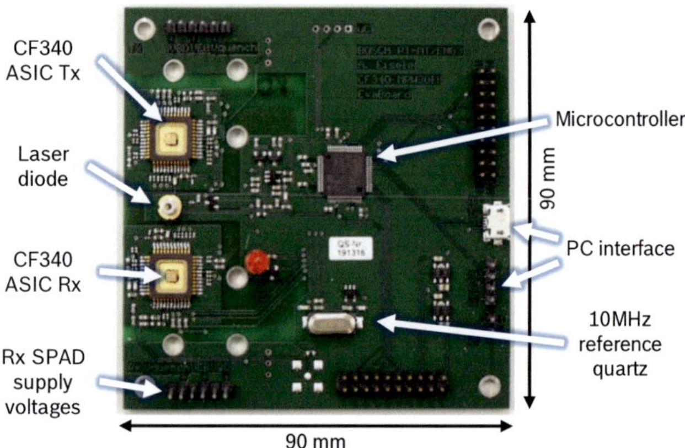

Fig. 6–17: Photograph of CF340 MPW2011 evaluation board

Fig. 6–18 presents a fully functional laser rangefinder prototype that has been realized in the course of this work. A collimator lens is mounted on top of laser diode. A low-cost plastic lens from a commercially available laser rangefinder [5] serves as the receiver lens. This lens features a rather sophisticated lens geometry comprising a 'near-range element' for projecting light onto the detector active area regardless of the target distance and thereby mitigates some of the effect of the parallax error (see patent application [111] for details of the lens geometry). The portion of the lens that projects light onto the detector at short distance is also referred to as near-range element. In a first order approximation, this lens can be thought of as a bifocal or multifocal lens analog to eyeglasses with special spatial zones for near range. Recall that this laser rangefinder is a non-imaging system that does not require a well-focused image.

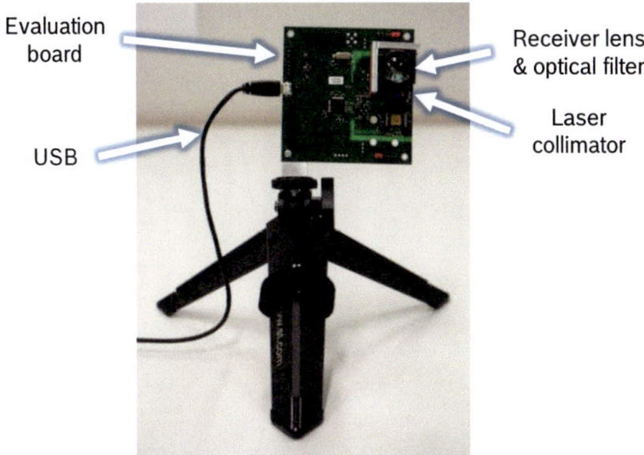

Fig. 6–18: Photograph of laser rangefinder prototype

6.2.2 Binning and Bin Homogenization

An improved binning architecture including bin homogenization is implemented in the CF340 MPW2011. Modifications to the binning architecture successfully reduce the pulse gaps that are strongly present in the previous implementation (see Fig. 6–7 for comparison). Moreover, careful device matching, including parasitic extraction in the design process, is expected to provide a more equalized bin width distribution.

6.2.2.1 Binning

Fig. 6–19 shows the absolute bin width for three samples of the CF340 MPW2011 (ASIC 2) in comparison to a sample from the previous CF340 MPW2009 (ASIC 1). Bin widths are again measured as described in Section 6.1.3.1. Tx and Rx phase shifters are set to a fixed state, i.e., bin homogenization is temporarily disabled. In spite of the improvements there is a significant bin width difference between different samples from the current ASIC. Non-idealities that are due to device mismatch in the semiconductor process cannot be eliminated by design.

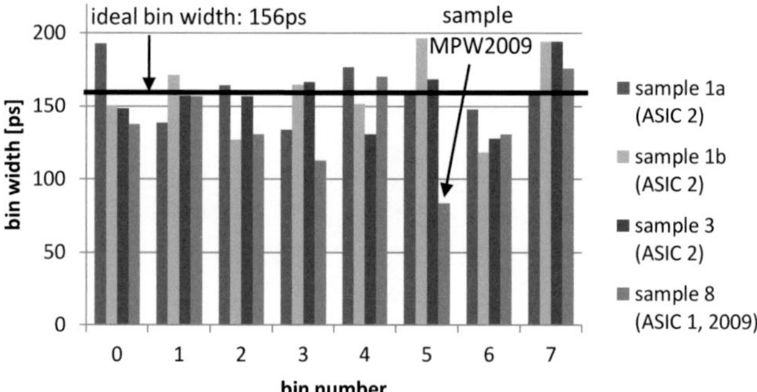

Fig. 6–19: Absolute bin widths. Samples 1a, 1b, 3 are CF340 MPW2011(ASIC 2) samples with reduced pulse gap. Sample 8 is a CF340 MPW2009 (ASIC 1) sample for comparison. Measurement at 800MHz modulation frequency.

6.2.2.2 Bin Homogenization

Two measurements without and with bin homogenization are presented in Fig. 6–20.

Fig. 6–20 (a) shows the non-ideal binning measured with sample 1a. Both Tx and Rx phase shifters in the sending and receiving path are set to a fixed state (sub0). This corresponds to a fixed relation between bin and counter. Hence, the relative bin width calculated from the counter values represents the bins of the non-ideal binning. The relative bin width is defined in Eq. (5.29).

In Fig. 6–20 (b) one has to distinguish between bins (e.g. red rectangle for bin 0) and counters (horizontal axis). The bottom row in the graph represents the first sub-measurement with counter values from Fig. 6–20 (a). During this first sub-measurement bin 0 increments counter 0, bin 1 increments counter 1 and so on. The second row represents the counter values at the second sub-measurement sub1. During this second sub-measurement bin 0 increments counter 7, bin 1 increments counter 0 and so on, i.e., each bin is assigned to the counter with the next lower index compared to the previous sub-measurement. This corresponds to a -2π/8 phase shift between sub-measurements. Further sub-measurements are performed analogously. Each bin increments each counter over the entire homogenization cycle. As a result the eight counters show even values despite the non-ideal binning.

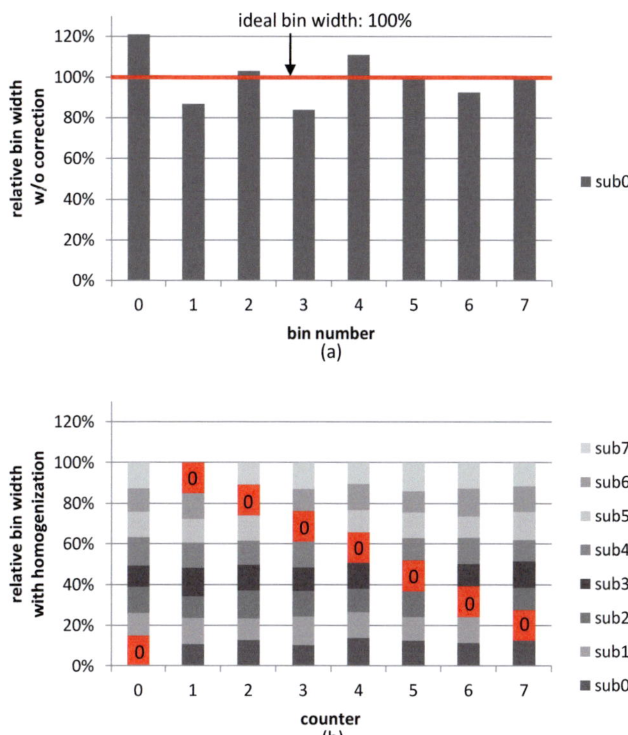

Fig. 6–20: Bin width homogenization. 800 MHz modulation frequency. (a) Normalized relative bin width without code density test (CDT) correction and without bin homogenization applied. Counts in each bin are normalized to a mean bin width of 100 %. (b) Normalized relative bin width with homogenization. Each row corresponds to a sub-measurement (sub0…sub7) with corresponding state of the phase shifters. The red elements mark the portion in each counter that is measured with bin 1. The counts of all sub-measurements are accumulated in the counters and then normalized to a mean bin width of 100 %.

There is a residual mismatch for the measured bin width after homogenization compared to the ideal 100 % bin width. At $N = 100 \times 10^6$ counts per counter, the measurement bin width deviates by 0.01 % from the ideal value. This corresponds to the relative photon noise $1/\sqrt{\overline{N}}$ expected for this average number of Poisson distributed events (see Section 3.2.2, Eq. (3.13)).

6.2.2.3 Crosstalk

Bin homogenization works perfectly well for temporally constant background light. However, crosstalk from the laser modulator to the binning circuits – both electrical and optical – is not corrected by bin homogenization. As mentioned previously, the CF340 MPW2011 (ASIC 2) is based on the same framework as the previous test chip CF340 MPW2009 (ASIC 1) with similar crosstalk sources present.

Fig. 6–21 shows a measurement of electrical crosstalk with one single ASIC. With bin homogenization, the bins are perfectly equalized when the on-chip laser modulator is turned off. The other curves show bin widths measured with bin homogenization at different, increasing laser modulator currents. The crosstalk from modulator to binning gives a periodic error on the fundamental modulation frequency.

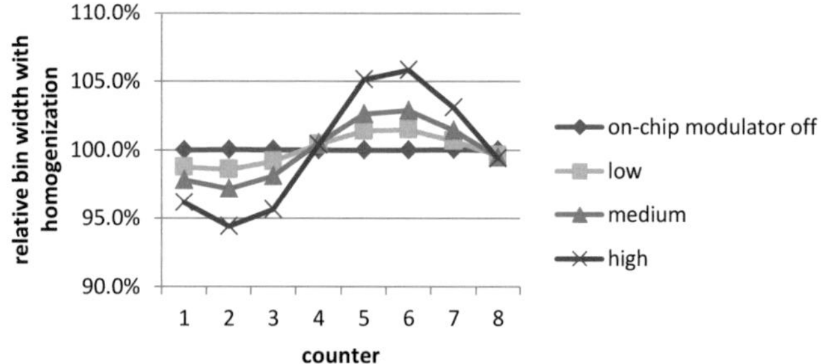

Fig. 6–21: Crosstalk measurement with single ASIC. Counter values are recorded at different laser modulator currents (off, low, medium, high) set in the laser modulator on the same ASIC. 800 MHz modulation frequency. The measured modulation depth on the fundamental frequency for modulator off, low, medium, and high current setting is 0.001 %, 1.51 %, 2.92 % and 5.79 %.

These measurements favorably compare with results from the CF340 MPW2009 (see Section 6.1.3.4, Fig. 6–11). The crosstalk amplitude at medium current is reduced from over 50 % in the previous ASIC to about 3 %. Even though the same framework was used, including the parallel bond wires of laser driver and SPAD supply, the crosstalk could be reduced by an improved design of the ASIC (laser driver shielded from binning and SPADs) and additional stabilizing capacitors on the evaluation board.

For measurement scenarios with low signal and strong background light, i.e., low signal modulation depth (see Fig. 5–22), the measured crosstalk can still be the dominant source of error. ASIC and housing have not been fully optimized for crosstalk yet. The evaluation board therefore features two independent ASICs, one of which can be used for driving the laser with its on-chip modulator turned on (Tx ASIC) and the second for detection with its on-chip modulator turned off (Rx ASIC). The two ASICs with the laser diode in-between are shown in the photograph in Fig. 6–22. The bin homogenization operates the Tx and Rx phase-shifters of both ASICs in parallel. Both chips are synchronized with the reference oscillator.

Fig. 6–22: Photograph of two CF340 MPW2011 test chips. The left ASIC modulates the laser (Tx ASIC), the right ASIC acts as the receiver (Rx ASIC) with its laser modulator turned off to prevent crosstalk from laser modulator to binning and SPADs. Both chips are synchronized with a common reference oscillator.

Using two separate but identical CF340 MPW2011 (ASIC 2) on one evaluation board reduces crosstalk by another order of magnitude. Counter values are recorded with the binning scheme on the Rx ASIC for different laser modulator currents (off / on at medium level) set in the laser modulator on the Tx ASIC (Fig. 6–23). When the Tx and Rx laser-modulator are turned off, the counter values measured with the Rx ASIC are – apart from photon noise – equal for all counters. When the Tx ASIC modulates the laser with medium current, there remains an apparent modulation depth of 0.25 % in the Rx binning. Some residual crosstalk is present because both ASICs and the laser diode share several electrical connections on the evaluation board. The common ground plane for example offers a crosstalk path. As a consequence, we expect a periodic distance error for measurement scenarios at low modulation depth.

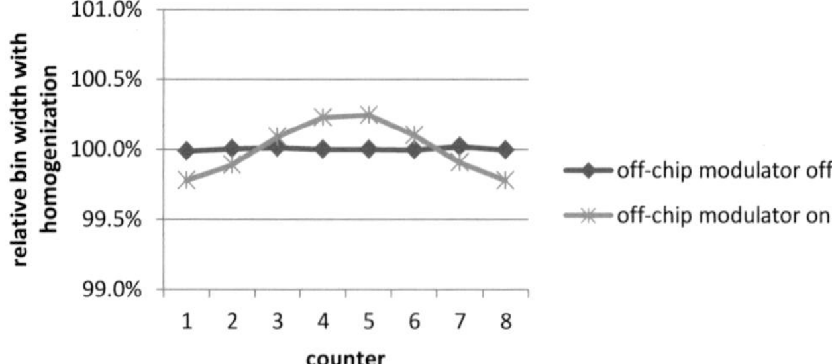

Fig. 6–23: Crosstalk measurement with two separate but identical chips on one evaluation board. Counter values are recorded with the binning scheme on the Rx ASIC for different laser modulator currents (off, on at medium level) set in the laser modulator on the Tx ASIC. 800 MHz modulation frequency. The modulation depth on the fundamental modulation frequency measured with the Rx ASIC is 0.001 % for Tx modulator of the Tx ASIC turned off and 0.25 % for Tx modulator of the Tx ASIC turned on.

6.2.3 Distance Measurement

Distance measurements with the CF340 MPW2011 presented in this section use the same measurement setup as the previous measurements (see Section 6.1.4). The transmitter is set to an average optical output power of 1 mW for laser class II eye safe operation. Burst modulation (see Section 5.2.3.2, Fig. 5–8 on page 137) has not been used. Unless otherwise stated we use a type 3 SPAD (see Section 4.2.1) in negative-drive option (see Section 3.4.4) with an active quenching circuit (see Section 3.4.2) set to 10 ns dead time.

A characterization measurement of the SPAD array shows that there are several picoseconds difference in response time between neighboring SPADs. Therefore, measurements with different SPADs would yield slightly different distances. Moreover, the photon detection efficiency varies slightly between SPADs and thus gives different count rates. A distance measurement that aims at evaluating the binning architecture rather than the study of light levels on the detector plane should therefore be conducted with one single SPAD. The image of the laser spot on the target wanders over the array because of parallax (see Section 5.2.4.2). In order to avoid switching between different SPADs, defocus is introduced to widen the image of the laser spot in the detector plane. The detector plane is moved slightly towards the receiver lens. Defocus widens the image of the laser spot at the price of decreased signal intensity. However, we can now perform a distance measurement over the entire range of the linear stage with a single defined SPAD providing its output signal to the binning.

6.2.3.1 Measurement with Low Background Light and High Modulation Depth

A first distance measurement is performed with two separate Rx and Tx evaluation boards to prevent on-board crosstalk. Each evaluation board carries one CF340 MPW2011. An external reference frequency generator synchronizes Tx and Rx ASICs. This setup also uses the receiver lens with near-range element described above.

At low background light, the modulation depth exceeds 70 % over the entire distance measurement range. This value demonstrates an advantage of this SPAD-based system over modulated CCDs in [32], where the modulation depth is attenuated by 3 dB already at a modulation frequency of 20 MHz. The present SPAD-based system operates at a modulation frequency of 800 MHz. A high modulation frequency is advantageous because the distance accuracy improves with an increasing modulation frequency (see Eq. (2.10)).

Fig. 6–24 shows three distance measurements (a) without bin width compensation (RAW); (b) with code density test correction (CDT; Section 5.3.4); and (c) with bin homogenization (HOM; Section 5.3.5). In 2 mm distance increments, a single-shot distance measurement is performed. Each data point represents a single measurement, including noise. The measurement time per distance is 0.8 s. A calibration measurement of 10 s time with non-modulated background light determines the bin width for CDT-correction before the actual distance measurement. At the far left end of the curves there is a systematic distance error common to all three measurements. This is due to the warm-up of the laser diode. This behavior has been verified in separate experiments. This drift effect is no longer present in a laser rangefinder with simultaneous distance and reference measurement derived from the same source. However, our experiment evaluates one single SPAD only (no reference path) and this does not allow any drift correction.

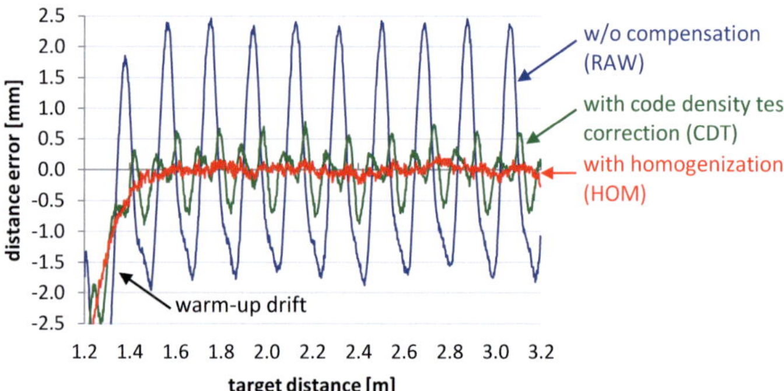

Fig. 6–24: Distance measurement. Distance error measured without bin with compensation (RAW, blue curve) using bin width correction via code density test (CDT, green curve) and with the proposed bin homogenization scheme using Tx and Rx phase-shifter (HOM, red curve). The maximum distance error (after warm-up for target distances >1.6m) decreases from 2.4 mm (RAW) over 0.8 mm (CDT) down to 0.24 mm (HOM). 800 MHz modulation frequency, 0.8 s measurement time per distance, single-shot measurements without averaging.

Without bin width correction, there is a periodic distance error, whose periodicity corresponds to the modulation period. The modulation period is also the repetition periodicity of the disturbed bins. The maximum distance error with the binning of the CF340 MPW2011 is 2.4 mm (after warm-up). Please note the different scale of the vertical axis compared to Fig. 6–15 which shows the distance error with the MPW2009 on a scale from –10 to 10 mm.

With code density test correction (CDT) of the bin width the maximum distance error reduces to 0.8 mm (after warm-up). The CDT corrected distance shows a systematic error with contributions at once and twice the modulation period.

With bin homogenization, the performance simulation of Section 5.4 (page 186 ff.) predicts the best distance measurement accuracy. The measurement time of 0.8 s is split up into eight sub-measurements of 0.1 s each. The current system achieves a maximum distance error of 0.24 mm with homogenization, including noise and any systematic errors (after warm-up). To the best of our knowledge, this is the highest precision for a SPAD-based ranging system reported so far.

6.2.3.2 Measurement at High Count Rate and Medium Modulation Depth

Measurement scenarios under practical conditions usually include a considerable amount of background light. A measurement at medium modulation depth and high count rate corresponds to a distance measurement at short distance and strong background light.

The laser rangefinder prototype (Fig. 6–18) which includes Tx and Rx ASIC on one evaluation board is used in this experiment. Distance is measured in 10 mm steps. At each distance, 100 repeated measurements are recorded to evaluate systematic errors from the mean measured value and noise from the fluctuation. Moreover, the measurement time is reduced to 80 ms for a more prominent noise contribution. Thus the averaging time for the mean distance error is $100 \cdot 80\,\text{ms} = 8\,\text{s}$. A waiting time is introduced between switching on the laser diode and starting the automated measurement sequence to avoid warm-up drift. The modulation frequency is 800 MHz.

Fig. 6–25 shows the measured count rate versus the target distance. The variation in intensity has three main contributions, namely the image of the (defocused) laser spot wandering over the detector plane due to parallax; light distribution characteristics of the receiver lens including near range element (see end of Section 6.2.1; [111]); and decreasing signal light over distance. The peak intensity lies between 1.4 m and 1.6 m distance. For a perfectly focused laser spot count rates would be higher but the spot would more quickly wander off the pixel. Background light (BL) is at a constant level for all target distances.

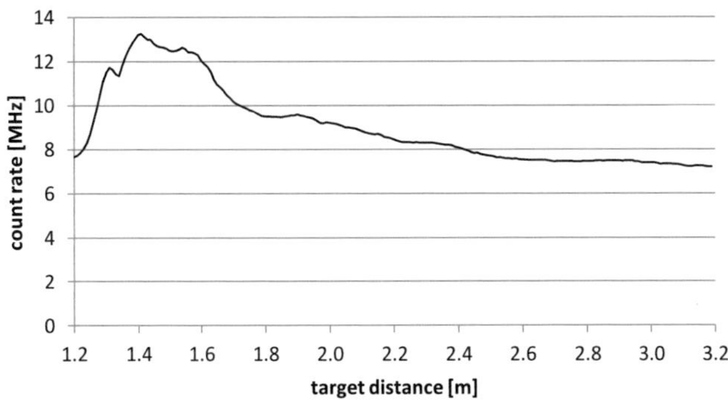

Fig. 6–25: Count rate measurement with laser rangefinder prototype.

When there is more signal light, the apparent modulation depth (Eq. (5.18)) measured at the receiver is higher. With CDT-correction or bin width homogenization the modulation depth follows the measured count rate (Fig. 6–26). Without bin width correction a variation of the modulation depth can be observed that has the same periodicity as the modulation frequency. This is because of an error phasor $\vec{E}$ from the uncorrected binning on top of signal phasor $\vec{S}$ (see phasor diagram in Fig. 5–2). A local maximum in modulation depth appears when both signal and error phasor point in the same direction. Analogously there is a local minimum, when both point in opposite direction.

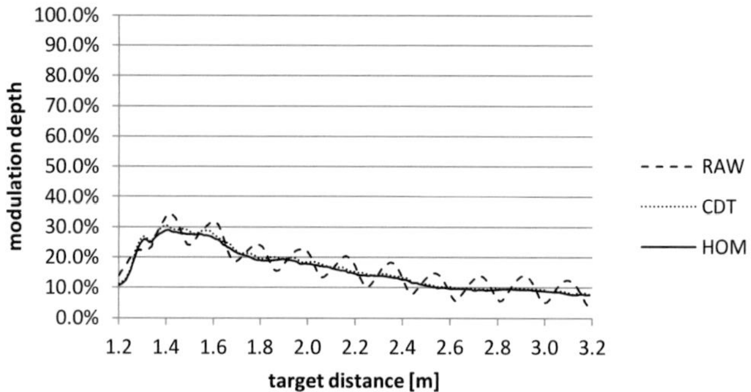

Fig. 6–26: Modulation depth with laser rangefinder prototype. Three curves are depicted: without bin width compensation (RAW), with code density test (CDT) and with homogenization (HOM).

Fig. 6–27 shows the mean distance error without bin width correction as a dashed line. As the modulation depth of the received signal decreases, the error contribution from the binning increases. In the two cases mentioned above, where the error phasor from the uncorrected binning is in phase or out of phase by π, the distance error from the binning is zero. This is the best-case binning error. The maximum error occurs, when the measured phasor and the error phasor are orthogonal to each other. This worst case, calculated from bin width non-idealities and the modulation depth above, is plotted as a grey solid line in Fig. 6–27. The calculated error nicely matches the maxima of distance errors measured without bin width correction in Fig. 6–27 over the entire range.

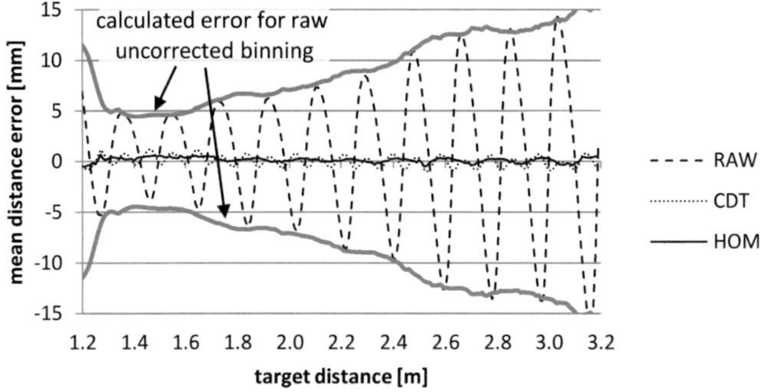

Fig. 6–27: Distance measurement with laser rangefinder prototype. Mean distance error measured without bin with compensation (RAW) using bin width correction via code density test (CDT) and with bin homogenization (HOM). The grey solid line shows the calculated maximum distance error expected without bin width correction applied. 800 MHz modulation frequency, mean distance error of 100 averaged measurements per distance, 80 ms measurement time per repetition.

Bin width correction lowers the mean distance error by about an order of magnitude. The results for CDT and homogenization are therefore depicted in the figure above for reference and plotted in Fig. 6–28 on a different scale.

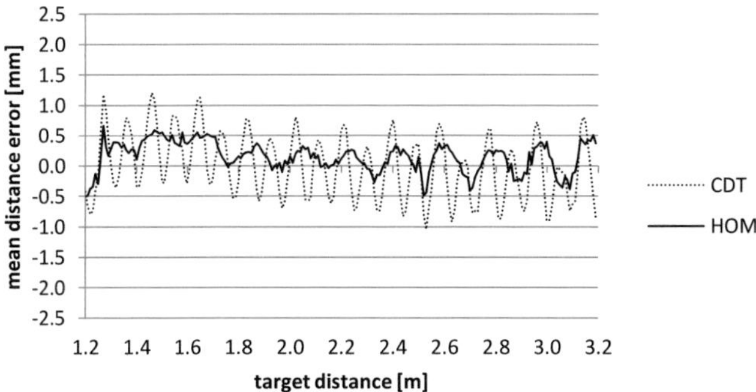

Fig. 6–28: Distance measurement with laser rangefinder prototype. Mean distance error measured with CDT-correction (CDT) and homogenization (HOM). 800 MHz modulation frequency, mean distance error of 100 averaged measurements per distance, 80 ms measurement time per repetition.

Using CDT, the maximum error is 1.2 mm with a predominantly double-periodic distance error that oscillates with ±0.6 mm amplitude. With bin homogenization, the maximum distance error reduces to 0.7 mm. A periodic oscillation of the distance error can be observed that increases with distance.

This can be attributed to the on-board crosstalk between Tx and Rx ASICs which has been measured in Section 6.2.2.3. As the signal modulation depth decreases from 1.4 m towards longer target distances, the crosstalk error increases.

Furthermore, there is a systematic error contribution common to both measurements in Fig. 6–28. We observe a distance offset which is strongest (around 0.4-0.5 mm) between 1.4 m and 1.6 m target distance. This coincides with the high count rate regime in Fig. 6–25. Although the link between count rate and distance offset is not obvious at a first glance, the distance offset is due to SPAD saturation / quenching effects (see Section 3.5.4), and is studied in a separate dedicated experiment.

Fig. 6–29 presents the standard deviation of the distance error of 100 repeated distance measurements. As expected, the repeatability error is low at high count rates and high modulation depth and high at low count rates and low modulation depth. Measurements with CDT-correction agree perfectly with the photon noise error predicted from theory (Section 5.3.2.4). Repeated measurements with bin width homogenization show a slightly higher statistical error that can be attributed to switching between different states in Tx and Rx path for the short measurement times in the present experiment.

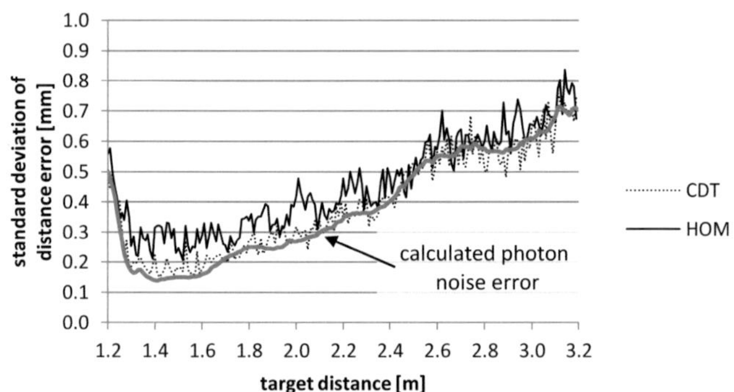

Fig. 6–29: Distance measurement with laser rangefinder prototype. Standard deviation of distance error measured with CDT-correction (CDT) and homogenization (HOM). The grey solid line shows the theoretical standard deviation calculated from the measured signal modulation depth and SPAD counts. 800 MHz modulation frequency, mean distance error of 100 averaged measurements per distance, 80 ms measurement time per repetition.

It should be noted that a measurement time of up to 4 s is allowed for the target system, which is a factor 50 longer than the experimental measurement time. As the measured counts scale linearly with measurement time t_{meas}, the relative error from photon noise scales with $1/\sqrt{t_{meas}}$. A 50 times longer measurement time therefore reduces the photon noise error by a factor $\sqrt{50} \approx 7$.

6.2.3.3 Measurement at Low Count Rate and Low Modulation Depth

Probably the most challenging operating condition for a laser rangefinder is a distant target exposed to strong background light. Because of range limitations in the laboratory, an optical attenuator in front of the receiver lens of the laser rangefinder prototype artificially reduces count rates and ensures the corresponding low signal count rates and low modulation depth.

Fig. 6–30 shows the measured modulation depth as a function of the target distance without bin width correction (RAW) and with bin width correction schemes code density test (CDT) and homogenization (HOM). With bin width correction, the measured modulation depth ranges between 2 and 10 %. Without bin width correction, the modulation depth drops to zero at about 2.1 m target distance. At this point the error phasor from the non-ideal binning becomes dominant over the signal phasor. A distance measurement without bin width correction is not possible beyond this point. Even though the measured modulation depth seems sufficiently strong at about 4.5%, this is not signal modulation. The constant error phasor from unequal bin widths is measured instead. This is nicely illustrated by the diagram in Fig. 6–31 where the measured phase fails to increase linearly with distance beyond 2.1 m target distance. The received modulated signal light is weaker than the impact of the non-ideal binning.

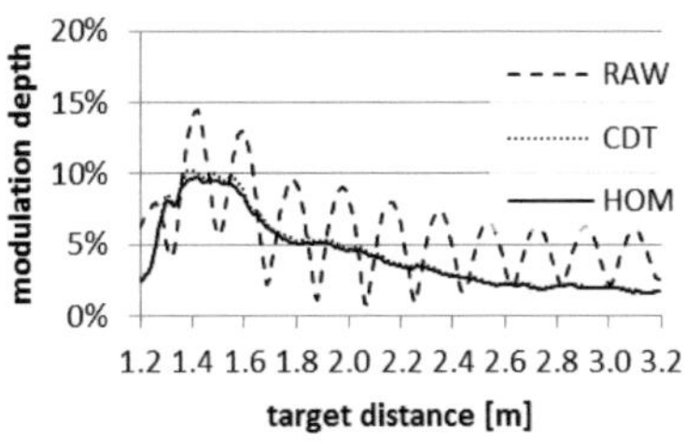

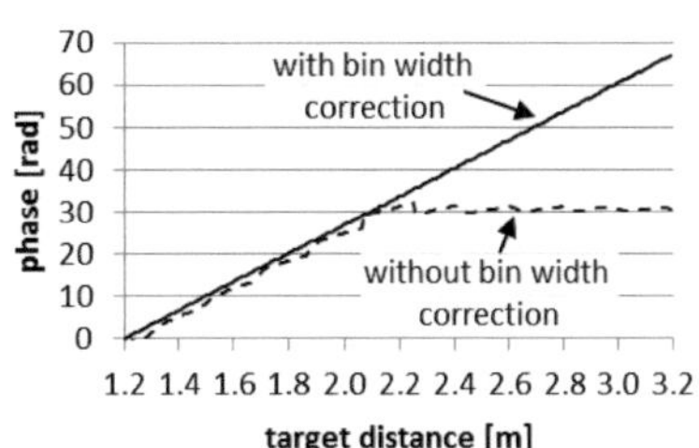

Fig. 6–30: Modulation during distance measurement with laser rangefinder prototype with attenuator in receiving path.

Fig. 6–31: Measured phase with and without bin width correction applied.

As the modulation depth decreases, the systematic error increases. On-board crosstalk now causes up to 2 mm systematic error. We have previously seen an advantage of bin homogenization over CDT-correction at higher modulation depth. However, this benefit diminishes towards a lower modulation depth, as the impact of crosstalk comes into play. A separate study of crosstalk effects on CDT-correction and homogenization is presented in Section 6.2.3.4 below.

In the previous experiment a 0.5 mm distance offset is observed between 1.4 m and 1.6 m distance. In the present measurement and at lower count rates the offset has vanished. This indicates that the distance offset is a count rate dependent SPAD saturation / quenching effect.

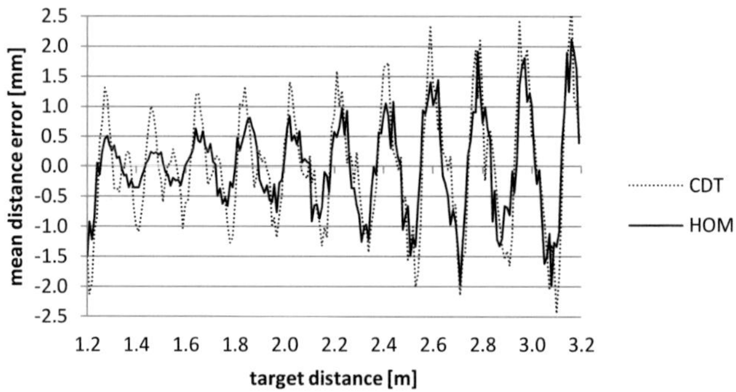

Fig. 6–32: Distance measurement with laser rangefinder prototype with attenuator in the receiving path. Mean distance error measured with CDT-correction (CDT) and homogenization (HOM). 800 MHz modulation frequency, mean distance error of 100 averaged measurements per distance, 80 ms measurement time per repetition.

6.2.3.4 Crosstalk Influence on Bin-Homogenization and CDT

Fig. 6–33 (a) shows a distance measurement with the laser rangefinder prototype at low background light and medium modulation depth. Recall that the on-chip laser modulator of the Tx ASIC is turned on and modulates the laser diode, whereas the on-chip laser modulator of the Rx ASIC is turned off. Under these conditions a distance measurement with bin width homogenization is superior to a distance measurement with CDT-correction.

Fig. 6–33 (b) shows a distance measurement with the same measurement setup but now the on-chip laser modulator of the Rx ASIC is turned on. The output signal of laser modulator of the Rx ASIC has the same external circuit components as the Tx ASIC, only the laser diode behind the coupling capacitor of the bias tee is replaced with a resistive load.

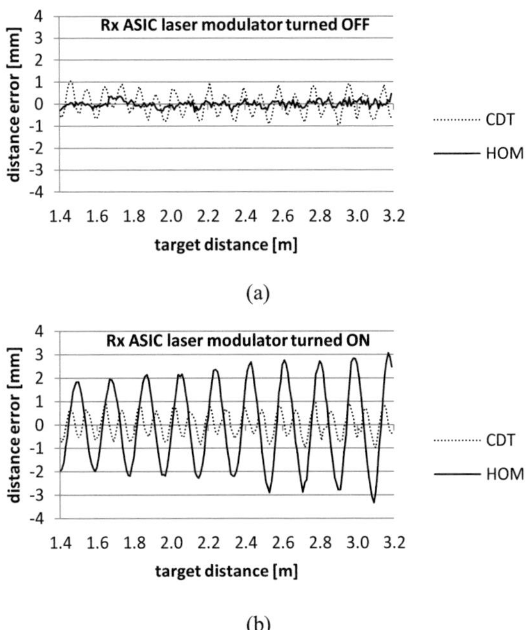

(a)

(b)

Fig. 6–33: Distance measurement with laser rangefinder prototype. Distance error after bin width correction using code density test (CDT) and bin width homogenization (HOM). (a) Tx ASIC laser modulator driving the laser diode while Rx ASIC laser modulator is turned off. (b) Tx ASIC laser modulator driving the laser diode while Rx ASIC laser modulator is turned on. 800 MHz modulation frequency, 0.8 s measurement time per distance, single-shot measurements without averaging.

Comparing the distance error with bin homogenization (HOM) in Fig. 6–33 (a) and (b) shows a strong periodic distance error with the same periodicity as the laser modulator when the on-chip Rx ASIC laser modulator is turned on in Fig. 6–33 (b). This behavior has been predicted in Section 5.3.6 because the crosstalk from laser driver onto the binning changes between sub-measurements.

Comparing the distance error with code density test correction (CDT) in Fig. 6–33 (a) and (b) yields a comparable distance error with and without the on-chip Rx ASIC laser modulator turned on. As proposed in Section 5.3.6, bin widths are calibrated with background light while the on-chip laser modulator is already on but with the laser DC current below lasing threshold. Hence the crosstalk of the laser modulator onto the binning is already included in the calibration. In this particular case, CDT-correction is superior over bin homogenization.

6.3 Conclusion and Outlook

We have successfully demonstrated a millimeter-precision laser rangefinder using a low-cost photon counter.

The heart of this laser rangefinder is an application-specific integrated circuit (ASIC) that integrates single-photon avalanche diodes (SPADs) alongside with a laser modulator and timing circuitry. Distances are measured using the single-photon synchronous detection (SPSD) principle.

For millimeter-precision distance measurements, the time of flight of the laser light must be known with picosecond precision. Because the semiconductor technology of SPADs has a limited RF performance, there is no fast clock signal available for high-speed sampling. Therefore, the sampling windows or 'bins' of the timing circuit are derived from different delay stages of a ring oscillator. Because of mismatch in the delay stages, the generated bins are not uniformly distributed over the modulation period, and are of unequal temporal width. Moreover, bin widths change from chip to chip and also with temperature and modulation frequency. Unequal bin widths cause a distance error that is especially problematic under strong background light (BL) or low signal modulation conditions. Hence bin width correction is mandatory.

The first investigated correction scheme is a code density test (CDT). Bin widths are measured with high precision prior to the actual distance measurement. CDT successfully eliminates distance errors due to background light, however, at the price of an additional calibration measurement.

A more promising second scheme as proposed in this work is bin homogenization. The association of sampling bin for capturing the measured SPAD pulses and the counter for counting the SPAD pulses can be chosen. We rotate the association of bin and counter during the measurement, so that the SPAD pulses in each non-ideal bin increment each counter alike. Thereby the background light is homogeneously distributed over all counters.

This bin homogenization has an advantage over CDT as it does not require any calibration measurement. Under ideal conditions, the distance error with bin homogenization is as low as 0.24 mm including noise and systematic errors.

This matches the requirement of a handheld laser rangefinder, where a low distance error at short distances is crucial. For a target a few meters away, a few millimeters of distance error can be tolerated.

In the presence of crosstalk, however, CDT-correction has an advantage over bin homogenization. It is proposed to modulate the laser below lasing threshold during CDT calibration. Hence, any electrical crosstalk influence is already included in the calibration. Bin homogenization is more susceptible to electrical crosstalk, because crosstalk slightly changes the binning between sub-measurements so that the background light is not homogeneously distributed over all counters. The present ASIC is not optimized for low crosstalk yet, and even after optimization there can be residual crosstalk. At low signal modulation depth and strong background light CDT can be the better choice.

Hence, CDT-correction is suggested for long distance measurements at low modulation depth, whereas bin homogenization is the better choice for short distance measurements at high modulation depths.

While the laser rangefinder prototype showed extraordinary distance measurement performance, there are still several technical challenges to be addressed.

On an industrial scale, nobody has ever tried to manufacture SPADs, not to speak of a SPAD-based laser rangefinder. Current test chips are individually processed multi-project wafer runs. SPADs have not yet reached a maturity level that ensures reliable operation. This process requires corner lots in semiconductor manufacturing (a design-of-experiments technique that refers to a variation of fabrication parameters), and extensive testing over the next years before a system-on-chip laser rangefinder can be industrialized. In conventional rangefinders, with discrete avalanche photodiodes, a defective photodetector can be sorted out, whereas in a SPAD-based rangefinder the entire ASIC must be rejected if SPADs are defective. Hence, if SPAD manufacturing does not reach a sufficient yield, the cost-benefit of a SPAD-based system over conventional rangefinders quickly ceases to exist.

Sophisticated simulation models are available for standard devices (transistors, diodes, etc.) in the current technology but not for SPADs. In particular, picosecond timing mismatch has been of limited interest for previous applications of the fabrication process. The laser rangefinder performance estimation is based in part on values extrapolated from the past experience of the designers. Future experiments will refine the input parameters of the performance estimation, and the outcome will certainly have to be altered in one direction or another.

Device mismatch is also crucial for the timing circuitry. The experiments carried out in the course of this work show very promising distance measurements under laboratory conditions. For a laser rangefinder to be employed in the rough environment of a constructions site, a sufficient performance must be ensured over the entire device lifetime under all specified operating conditions. Calibration concepts for manufacturing are to be developed and field tests must be carried out.

The warm-up distance error after switching on the device must be corrected. As a solution, multiple SPADs can easily be integrated for performing a reference measurement on the same chip. The absolute temperature drift then reduces to a drift difference between target and reference SPAD array.

In order to address the SPAD saturation problem, the quenching and recharge circuit should be tailored for a specific SPAD. Moreover, the lower capacitance of a SPAD using the positive drive circuit option (Fig. 3–25, Fig. 4–34) promises to shorten the recharge time by a factor of 3.8. Hence the probability of a secondary pulse during active recharge reduces.

For a better background light tolerance, a narrower optical band-pass filter should be considered. Currently a 40 nm band-pass is employed. However, with dropping prices a narrowband filter is the easiest way to improve background light tolerance. It should be noted that the temperature-dependent wavelength drift of the laser diode has to be borne in mind when reducing the filter width.

Further, on-chip crosstalk remains one of the core problems. If crosstalk still persists, an on-chip laser modulator with low power can be combined with an external amplifier. This keeps driving currents inside the chip at a moderate level. Unfortunately, additional external components would be required causing additional cost.

The implemented ASICs serve as a platform for an experimental verification of individual system building blocks. The chip has to be optimized further in its entirety. For example, the ASIC comprises a distance measurement unit without reference path and has not been optimized for electrical crosstalk. Therefore, our experiments use separate, synchronized ASICs for the sending and for the receiving path. The system performance that can ultimately be reached can only be determined when measuring with an ASIC fabricated according to final system architecture.

The SNR can be further improved by burst modulation. In the case where background light dominates, the modulation depth can be quadrupled by choosing a 25% burst duty cycle (Fig. 5–8). Preliminary measurements showed that burst modulation is feasible with the low-cost laser diodes used in the present experimental setup.

The experiments presented in this work are designed to measure distance at a single modulation frequency. Measurements at multiple frequencies are required to cover the full unambiguity range corresponding to the system's minimum difference frequency. For the current test chips, the minimum difference frequency is 1.25 MHz, which corresponds to an artificial wavelength of 120 m, and thus theoretically enables distance measurements of up to 120 m with the ASIC CF340 MPW2011.

In conclusion, this work paves the way towards a millimeter-precision laser rangefinder at low cost. Since the experiments show very promising results, a commercial product can be expected in the near future.

Appendix I

This appendix provides further information on selected topics that have been referred to in the main part of the present work.

A. Definition of Measurement Accuracy

- The *distance measurement error* ΔL is the difference between the measured target distance L_{meas} and the true target distance L,

$$\Delta L = L_{\mathrm{meas}} - L. \tag{A.1}$$

The distance measurement accuracy of a laser rangefinder is impaired by systematic and non-systematic distance measurement errors. At a true target distance L, the mean measured target distance $\Lambda_L c$ is given by the mean value of N repeated measured target distances L_n,

$$\Lambda_L = \frac{1}{N} \sum_{n=1}^{N} L_n. \tag{A.2}$$

Using Eq. (A.2), the *systematic distance error* μ_L at a true target distance L is defined as the mean distance error of the N repeated distance measurements L_n,

$$\mu_L = \Lambda_L - L. \tag{A.3}$$

The variance σ_L^2 of the N repeated distance measurements L_n at a true target distance L is given by

$$\sigma_L^2 = \frac{1}{N-1} \sum_{n=1}^{N} \left(L_n - \Lambda_L\right)^2. \tag{A.4}$$

The standard deviation σ_L defines the *non-systematic distance error* σ_L. The non-systematic distance error is also referred to as the repeatability error.

Neither the systematic distance error $|\mu_L|$ nor the non-systematic distance error σ_L increase monotonic with L. Fig. A-1 shows a graph of the distance error as a function of the true target distance L. The *specified range-dependent accuracy* $\xi(L)$ is defined by the upper and lower boundaries, given by

$$\xi(L) = \pm\left(\xi_0 + \xi_1 L\right) \tag{A.5}$$

where ξ_0 is an offset, and ξ_1 is a slope in [mm/m] that accounts for an increase of the distance measurement error with target distance. As sketched in Fig. A-1, these boundaries are selected to contain at least the systematic distance error μ_L plus/minus twice the repeatability error σ_L at each true target distance L,

$$\left|\xi(L)\right| \geq \left|\mu_L\right| + 2\sigma_L . \tag{A.6}$$

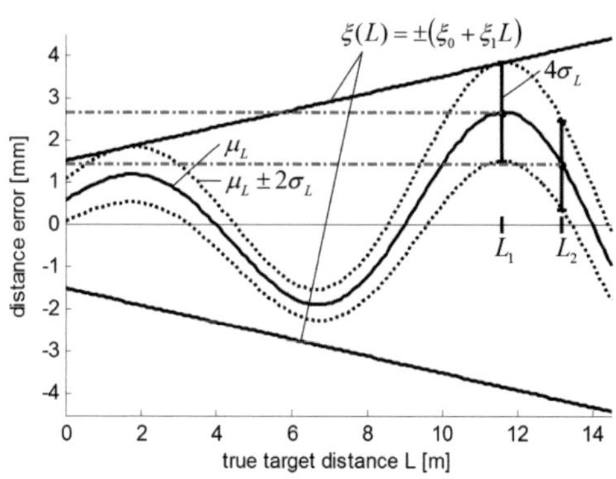

Fig. A-1: Distance error vs. true target distance. μ_L systematic distance error, σ_L non-systematic distance error, $\xi(L)$ upper and lower linear boundary curves.

For defining a *worst-case distance accuracy* $\varepsilon(L)$, a range of true target distances $l \leq L$ is regarded. The worst-case systematic distance error $\varepsilon_\mu(L)$ and the worst-case non-systematic distance error $\varepsilon_\sigma(L)$ then are

$$\varepsilon_\mu(L) = \max\left(\left|\mu_l\right|\right) \ \forall\, l \leq L \quad \text{and} \quad \varepsilon_\sigma(L) = \max\left(\sigma_l\right) \ \forall\, l \leq L_{\max} . \tag{A.7}$$

The worst-case distance accuracy $\varepsilon(L)$ occurs, when $\varepsilon_\mu(L)$ and $\varepsilon_\sigma(L)$ coincide,

$$\varepsilon(L) = \varepsilon_\mu(L) + 2\varepsilon_\sigma(L) . \tag{A.8}$$

The performance simulations of the laser rangefinder system in Chapter 5 typically simulate this worst-case distance accuracy $\varepsilon(L)$ in order to estimate an upper boundary of the distance error that can be expected. If $\varepsilon_\mu(L)$ and $\varepsilon_\sigma(L)$ do not coincide, the measurement accuracy will be even better. In Fig. A-1 the worst-case distance error occurs at target distance L_1, whereas at a larger distance L_2 the measurement error is lower again.

The *specified range-dependent accuracy* $\xi(L)$ is chosen such that it comprises the *worst-case distance accuracy* $\varepsilon(L)$ at any specified target distance.

B. Derivation Inter-Arrival Time Distribution

The inter-arrival time Δt is the time between two subsequent (detected) photons. As shown in Fig. 3–5 the output of a SPAD is a series of pulses that represent detected photons.

Using the Poisson distribution in Eq. (3.12), the probability $\mathrm{Prob}(N, p\Delta t)$ of detecting exactly N photons during a time interval Δt at a constant photon rate p (for the definition of the time-dependent photon rate $p(t)$ see Eq. (3.4)) is given by

$$\mathrm{Prob}(N, p\Delta t) = \frac{(p\Delta t)^N}{N!} \exp(-p\Delta t). \tag{B.1}$$

From Eq. (B.1) it follows that the probability of detecting no photon in a time interval Δt is

$$\mathrm{Prob}(0, p\Delta t) = \exp(-p\Delta t). \tag{B.2}$$

We are now interested in the inter-arrival time or more precisely the probability density distribution $f(\Delta t)$ of the inter-arrival time Δt between two successive photons. The probability of no photon arriving within a time interval Δt in Eq. (B.2), is equivalent to the probability that the time between two successive incident photons is larger than Δt. Thus the probability of no photon arriving within a time interval Δt is equal to the integral of the probability density over time interval from Δt and ∞ [131], which gives:

$$\text{Prob}(0, p\Delta t) = \int_{\Delta t}^{\infty} f(T)\,dT$$

$$\frac{d}{d\Delta t}\text{Prob}(0, p\Delta t) = \frac{d}{d\Delta t}\int_{\Delta t}^{\infty} f(T)\,dT$$

$$= \int_{\Delta t}^{\infty} \frac{\partial f(T)}{\partial \Delta t}\,dT + \underbrace{\frac{d\infty}{d\Delta t}}_{=0} f(\Delta t) - \underbrace{\frac{d\Delta t}{d\Delta t}}_{=0} f(\Delta t)$$

$$- p\exp(- p\Delta t) = -f(\Delta t).$$

(B.3)

The probability density for a time interval Δt between two successive photons is therefore an exponential decay,

$$f(\Delta t) = p\exp(- p\Delta t). \tag{B.4}$$

A short inter-arrival time between photons of a given photon energy corresponds to a high instantaneous power. Thus the exponential distribution of the probability density of the inter-arrival time can alternatively be described in terms of an exponential distribution of the probability density of the instantaneous power [42].

C. Sampling of Received Signal

This section provides further details about the sampling of a received signal and some considerations leading towards the bin homogenization scheme disclosed in Section 5.3.5.

Time-Continuous Periodic Signal

The received signal is assumed to be an ideal sinusoidal signal with offset A_0 and amplitude A_1. The angular modulation frequency is denoted by $\omega_m = 2\pi f_m$, the phase of the received signal by Φ_{Rx}. The received signal is thus given by

$$s_{Rx}(t) = A_0 + A_1 \cos(\omega_m t + \Phi_{Rx}). \tag{C.1}$$

The signal is periodic with a period $T = 1/f_m$. The information of the signal can thus represented by a single period. Additional periods do not contain further information.

The (time-continuous) periodic signal given in Eq. (C.1) can be represented by a Fourier series in the form of

$$s_{Rx}(t) = \sum_{l=-\infty}^{\infty} c_l e^{jl\omega_m t}. \tag{C.2}$$

In the frequency-domain, the time-continuous periodic signal is represented by the frequency-discrete components of the Fourier series from Eq. (C.2). The frequency components are located at integer multiples of the modulation frequency f_m.

The <u>time-continuous periodic signal</u> is thus represented by an infinite <u>frequency-discrete spectrum</u>.

Ideal Sampling

The time-continuous periodic signal is sampled at the receiver. The ideal sampling can be described as a multiplication of the receives signal with a Dirac comb representing the sampling function $a(t)$. The sampling rate is $f_a = 1/T_a$. The sampled signal is a time-discrete signal given by

$$s_a(t) = s_{Rx}(t) a(t) = s_{Rx}(t) T_a \sum_{k=-\infty}^{\infty} \delta(t - kT_a). \tag{C.3}$$

The spectrum of the sampled signal is given by

$$S_a(f) = S_{Rx}(f) * \sum_{k=-\infty}^{\infty} \delta(f - kf_a). \tag{C.4}$$

Due to the sampling process, the spectrum of the received signal $s_{Rx}(f)$ repeats periodically every f_a. To avoid overlap of neighboring representations of the spectrum, the spectrum must have a maximum width of $f_a/2$.

Advantageously, the sampling rate f_a is an integer multiple of the modulation frequency f_m of the periodic signal $s_{Rx}(t)$. This enables a cost-effective implementation because modulation and sampling can be derived from the same source conveniently.

Finite Sampling Windows

In practice, sampling with Dirac pulses of infinitesimal width is not possible. Sampling windows are used instead, for example rectangular sampling windows of finite width b.

The following definitions are used:

$$\text{rect}\left(\frac{t}{b}\right) = \begin{cases} 1/b, & |t| < b/2 \\ 0, & |t| > b/2 \end{cases}$$

$$\text{sinc}\,(bf) = \frac{\sin(\pi b f)}{\pi b f}.$$

(C.5)

The sampled signal is a time-discrete signal given by

$$s_a(t) = \text{rect}\left(\frac{t}{b}\right) * \left[s_{Rx}(t) T_a \sum_{k=-\infty}^{\infty} \delta(t - kT_a) \right].$$

(C.6)

The spectrum of the sampled signal is given by

$$S_a(f) = \text{sinc}\,(bf) S_{Rx}(f) * \sum_{k=-\infty}^{\infty} \delta(f - kf_a).$$

(C.7)

The term $\text{sinc}\,(bf)$, introduced by the rectangular sampling window, attenuates the amplitude of the spectrum and can cause a sign change. A sign change corresponds to a 180° phase shift of the detected signal. However, provided that sampling window width b is smaller than the period T, the sign does not change.

In order to make use of all the energy contained in the received signal, the sum of all sampling window widths should equal the period. For example, for eight sampling windows per period, the sampling window width is chosen as $b = T/8$. A phase shift must be avoided because the target distance is determined from the phase difference between sent and received signal (see Section 2.2.3.2).

Finite Sampling Windows of Different Widths and Non-Equidistant Distribution in Time

As described in detail in Section 5.3.3, the 'binning' of the laser rangefinder ASIC comprises eight sampling windows that are not necessarily equidistantly distributed in time and do not necessarily have the same width. These eight non-ideal sampling windows repeat periodically, wherein the period corresponds to the period $T = 1/f_m$ of the received signal.

The individual widths of the sampling windows are denoted by b'_k; the positions with respect to the first period are denoted by t'_k. The sampled signal is a time-discrete signal given by

$$s_a(t) = \sum_{k=0}^{7} \left(b'_k \, \mathrm{rect}\left(\frac{t}{b'_k}\right) * s_{Rx}(t) \, T \sum_{k=-\infty}^{\infty} \delta(t - kT - t'_k) \right). \tag{C.8}$$

Since the area of the rect-function in Eq. (C.5) is normalized to 1, the quantity b'_k has been introduced in Eq. (C.8) to account for the bin width.

D. Modulation Depths with Different Modulation Schemes

The laser rangefinder presented herein uses frequency-domain reflectometry to determine the target distance. The phase difference between sent and received signal at the fundamental modulation frequency is evaluated. For a good signal-to-noise ratio, a strong modulation amplitude of the signal is desired. However, eye safety, as a mandatory constraint, limits the average laser output power of the laser rangefinder to 1 mW [7].

The ratio of the amplitude of the signal at the fundamental modulation frequency to the average of the signal is referred to as the modulation depth or modulation index m_{signal}. The modulation depth depends on the selected modulation scheme as will be shown below.

Sinusoidal Signal

Fig. D-1 illustrates a purely sinusoidal modulation. The average value is 1. The signal is given by

$$y(x) = 1 + \sin x. \tag{D.1}$$

The maximum modulation depth is $m_{\mathrm{signal}} = 100\%$.

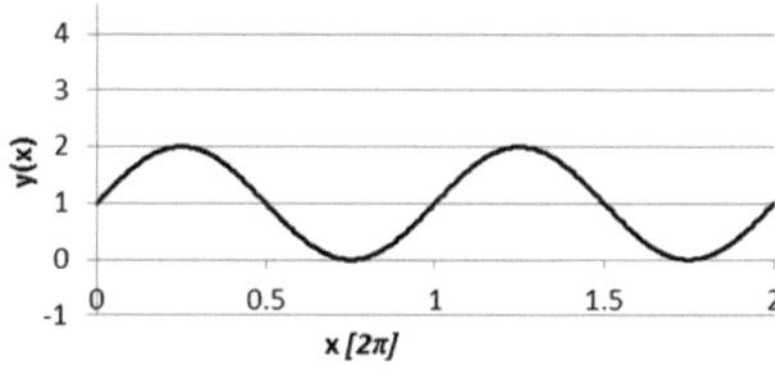

Fig. D-1: Sinusoidal intensity modulation. Modulation depth 100%.

Half Waves of Sinusoidal Signal

Fig. D-2 illustrates a signal comprising only positive half waves of a sinusoidal intensity modulation. The average value is 1. The signal can be described as a Fourier series given by

$$y(x) = 1 + \frac{\pi}{2}\sin x - 2\left(\frac{\cos 2x}{1 \cdot 3} + \frac{\cos 4x}{3 \cdot 5} + \frac{\cos 6x}{5 \cdot 7} + ...\right). \tag{D.2}$$

The maximum modulation depth is $m_{signal} = \pi/2 \approx 157\%$.

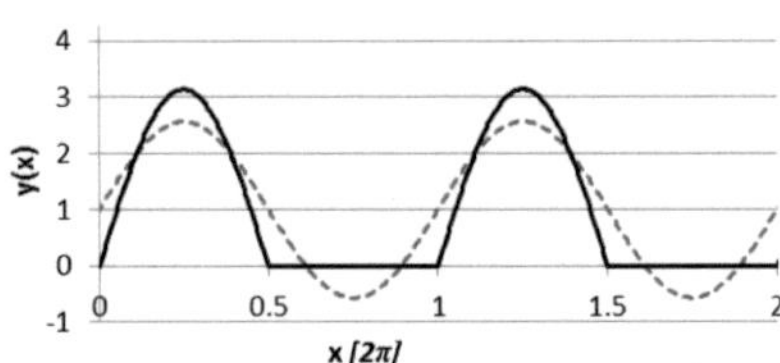

Fig. D-2: Signal comprising positive half waves of sinusoidal intensity modulation (solid line). Modulation depth of 157 % of at the fundamental modulation frequency (dashed line).

Pulse Wave with Rectangular Pulses of Duty Cycle 50 %

Fig. D-3 illustrates a rectangular signal with pulses of duty cycle 50 %. The average value is 1. The signal can be described as a Fourier series given by

$$y(x) = 1 + \frac{4}{\pi}\sin x + \frac{4}{\pi}\left(\frac{\sin 3x}{3} + \frac{\sin 5x}{5} + ...\right). \tag{D.3}$$

The maximum modulation depth is $m_{signal} = 4/\pi \approx 127\%$.

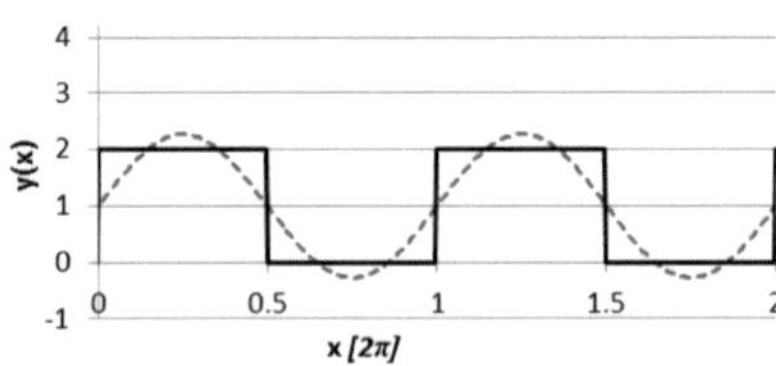

Fig. D-3: Pulse wave with rectangular pulses of duty cycle 50% duty cycle (solid line). Modulation depth of 127 % of at the fundamental modulation frequency (dashed line).

Pulse Wave with Rectangular Pulses of Duty Cycle D

In general and for a rectangular pulse, the modulation depth depends on the duty cycle D. Fig. D-4 illustrates an example of a rectangular signal with pulses of duty cycle 25 %. The average value is 1. The maximum modulation depth for a duty cycle of 25 % is $m_M \approx 180\%$.

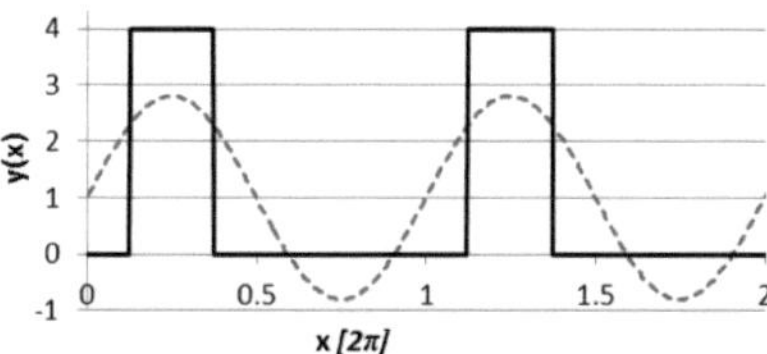

Fig. D-4: Pulse wave with rectangular pulses of duty cycle 25% duty cycle (solid line). Modulation depth of 180 % of at the fundamental modulation frequency (dashed line).

The signal can be defined as a Fourier series given by

$$y(x,D)=1+\frac{1}{D}\sum_{n=1}^{\infty}\frac{2}{n\pi}\sin(\pi nD)\cos(nx)$$
$$=1+2\frac{\sin(\pi D)}{\pi D}\cos x+2\frac{1}{\pi D}\sum_{n=2}^{\infty}\frac{1}{n}\sin(\pi nD)\cos(nx). \tag{D.4}$$

The dashed line in Fig. D-5 illustrates the signal for a Fourier series for $n \leq 6$. The higher frequency components contribute to a non-negative signal. An intensity modulated signal is always non-negative.

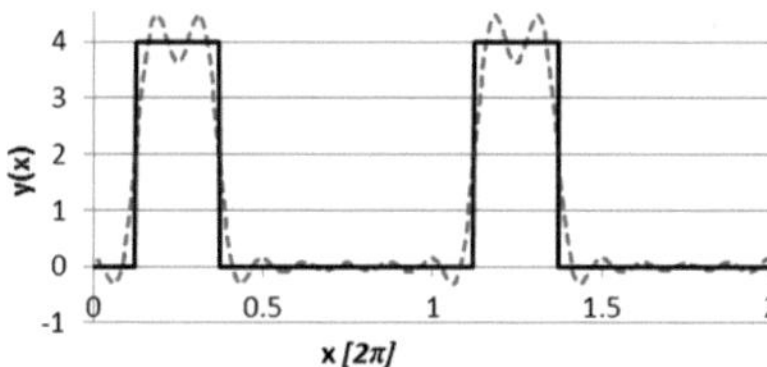

Fig. D-5: Pulse wave with rectangular pulses of duty cycle 25% duty cycle (solid line). Signal for Fourier series up to six times the fundamental modulation frequency (dashed line)

Using the definition for the sinc-function of Eq. (C.5), the Fourier series of Eq. (D.4) can be rewritten as

$$y(x,D) = 1 + 2\operatorname{sinc}(D)\cos x + 2\frac{1}{\pi D}\sum_{n=2}^{\infty}\frac{1}{n}\sin(\pi n D)\cos(nx). \qquad\text{(D.5)}$$

The modulation depth can thus be written as a function of the duty cycle, $m_{signal}(D) = 2\operatorname{sinc}(D)$. For a duty cycle of 25%, the modulation depth is thus $m_{signal} = 2\operatorname{sinc}(25\%) \approx 180\%$. Further decreasing the duty cycle gives a maximum modulation depth of 200%.

Limitations

As shown above, for a signal of limited average power, the modulation depth can be increased to more than 100 % by selecting an appropriate modulation scheme. However, there exist limitations.

Firstly, the hardware of the transmitter has to be capable of generating the desired signal.

Furthermore, care has to be taken that the signal can be properly reconstructed at the receiver. The spectrum can comprise frequency components at frequencies higher than the Nyquist frequency that can give rise to aliasing (for aliasing see Fig. 5–20 on page 155 and Appendix C above).

Glossary

Constants

c	Speed of light	299 792 458 m/s
π	Pi	3.142
h	Planck constant	6.626×10^{-34} Js
$\hbar = \dfrac{h}{2\pi}$	Reduced Planck constant	1.055×10^{-34} Js
k	Boltzmann constant	$1.381 \times 10^{-23} J/K$

Acronyms

APD	Avalanche Photodiode
APP	Afterpulse Probability
ASIC	Application-Specific Integrated Circuit
BIN	Sampling window for sampling a received signal
BINNING	Sampling architecture for sampling windows in time
BL	Background Light
CCD	Charge-Coupled Device
CDT	Code Density Test
CMOS	Complementary Metal Oxide Semiconductor
CP	Charge Pump
CQFP	Ceramic Quad Flat Package (semiconductor chip package)
DCR	Dark Count Rate
DIY	Do-It-Yourself, non-professional craftsman
DLE	Model name of Bosch laser rangefinder
DNL	Differential Nonlinearity
DSP	Digital Signal Processor

DT	Dead Time
FDR	Frequency-Domain Reflectometry
FF	Fill Factor
FFT	Fast Fourier Transform
FWHM	Full Width at Half Maximum
G-APD	Geiger Mode Avalanche Photodiode (=SPAD)
GLM	Model name of Bosch laser rangefinder
HMI	Human Machine Interface
HOM	Bin homogenization
i^{2}c	Serial data bus for communication with ASIC
INL	Integral Nonlinearity
IP54	International Protection rating number 54 according to IEC 60529, defines the degree of protection against dust, water, mechanic shock
JT	Jitter
LD	Laser Diode
LED	Light Emitting Diode
LF	Loop Filter
LRF	Laser Rangefinder
LQFP	Low-profile Quad Flat Package (semiconductor chip package)
MCU, µC	Microcontroller
MLR	Multi-Layer Reticle
MPW	Multi-Project Wafer
PCB	Printed Circuit Board
PDE	Photon Detection Efficiency
PFD	Phase Frequency Detector
PLL	Phase-Locked Loop

PLR	Model name of Bosch laser rangefinder
PMD	Photonic Mixer Device
PP	Pixel Pitch
RF	Radio Frequency
RMS	Root Mean Square
ROI	Region of Interest
SLR	Single-Lens Reflex (type of camera)
SNR	Signal-to-Noise Ratio
SONAR	SOund Navigation And Ranging
SPAD	Single-Photon Avalanche Diode
ST	STMicroelectronics
TCSPC	Time-Correlated Single-Photon Counting
TDR	Time-Domain Reflectometry
TOF	Time of Flight
USB	Universal Serial Bus
VCO	Voltage-Controlled Oscillator

Symbols

V_{bd}	Breakdown voltage
V_{eb}	Excess bias voltage
V_{th}	Threshold voltage
m_{signal}	Modulation depth of signal
m_M	Modulation depth determined from counts
m	Ideal count rate
n	(Actually measured) Count rate
p	Photon rate
P_{opt}	Optical power

Prob_{AP}	Afterpulse probability
W	Work, Energy
λ_0	Optical wavelength, typ. $\lambda_0 = 635\,\mathrm{nm}$
f_0	Optical frequency and $\omega_0 = 2\pi f_0$
λ_m	Modulation wavelength
f_m	Modulation frequency and $\omega_m = 2\pi f_m c$
μ	Mean value
σ	Standard deviation
σ^2	Variance
N_x, $\#x$	Number of x, for example a number of photons
τ	Time of flight
$\Delta\tau$	Time measurement error
L	Target distance
ΔL	Distance measurement error
φ	Phase difference between sent and received signal
$\Delta\varphi$	Phase measurement error
Φ_{Tx}/Φ_{Rx}	Phase of sent/received signal
k	Bin number

Bibliography

[1]   "History of length measurement," http://www.npl.co.uk/educate-explore/factsheets/history-of-length-measurement/

[2]   "Resolution 1: definition of the metre," 17th General Conference on Weights and Measures, 1983

[3]   Wikipedia, "Distance measurement," http://de.wikipedia.org/wiki/Entfernungs-messung, 29.08.2012

[4]   GLM 250 Manual, Robert Bosch GmbH, Leinfelden-Echterdingen, 2009.

[5]   GLM 80 Manual, Robert Bosch GmbH, Leinfelden-Echterdingen, 2011.

[6]   Wikipedia, "Lux", http://de.wikipedia.org/wiki/Lux_(Einheit), 08.06.2011

[7]   Safety of Laser Products: IEC-60825-1:2007, 2007.

[8]   R.A. Jarvis, "A perspective on range finding techniques for computer vision," IEEE Trans. on Pattern Analysis and Machine Intelligence, vol. PAMI-5, 122-139, 1983

[9]   C. Niclass, "Single-photon image sensors in CMOS: picosecond resolution for three-dimensional imaging," PhD Thesis no. 4161, École Polytechnique Fédérale de Lausanne, Switzerland, 2008

[10]   The PrimeSensor Reference Design 1.08, PrimeSense Israel, Tel-Aviv, 2010.

[11]   G. Wiora, "Sonar principle," http://commons.wikimedia.org/wiki/File:Sonar_Principle_EN.svg, 2005

[12]   W. Freude, "Messverfahren der optischen Nachrichtentechnik: Lichtwellenleiter," Institute of Photonics and Quantum Electronics, Karlsruhe Institute of Technology, 1989

[13]   W. Eickhoff, R. Ulrich, "Optical frequency domain reflectometry in single-mode fiber," Appl. Phys. Lett., 39, 693-695, 1981

[14]   R.I. MacDonald, "Frequency domain optical reflectometer," Appl. Opt., 20, 1840-1844, 1981

[15]   G.P. Agrawal, "Fiber-optic communication systems", 2nd edition, Willey-Interscience, New-York, 1997

[16]   E. Goulielmakis, M. Schultze, M. Hofstetter, V. S. Yakovlev, J. Gagnon, M. Uiberacker, A. L. Aquila, E. M. Gullikson, D. T. Attwood, R. Kienberger, F. Krausz, U. Kleineberg, "Single-cycle nonlinear optics," Science, 320, 1614-1617, 2008

[17]   R. Dändliker, Y. Salvadé, E. Zimmermann, "Distance measurement by multiple-wavelength interferometry," J. Opt., 29, 105, 1998

[18]   M. Hugenschmidt, "Laser-Entfernungsmessung," in Lasermesstechnik, Diagnostik der Kurzzeitphysik, Springer, Heidelberg, 109-130, 2007

[19]   A. Eisele, O. Wolst, B. Schmidtke, U. Skultety-Betz, "Optische Abschirm-Vorrichtung zum Trennen von optischen Pfaden", German patent application DE102010041937A1, 2012

[20]   A. Eisele, B. Schmidtke, R. Schnitzer, "Messvorrichtung zur Messung einer Entfernung zwischen der Messvorrichtung und einem Zielobjekt mit Hilfe optischer Messstrahlung," German patent application DE102011005740A1, 2012

[21]   B. Schmidtke, O. Wolst, A. Eisele, „Optical Measuring Device," PCT patent application WO002011029645A1, 2011

[22]   B. Schmidtke, O. Wolst, A. Eisele, "Optical Measuring Device," PCT patent application WO002011029651A1, 2011

[23]   E. R. Moutaye, H. Tap-Beteille, "CMOS avalanche photodiode embedded in a phase-shift laser rangefinder," IEEE Trans. Electron Devices, 55, 3396-3401, 2008

[24]   R. Lange, P. Seitz, A. Biber, R. Schwarte, "Time-of-flight range imaging with a custom solid state image sensor," Proc. SPIE 3823, Laser Metrology and Inspection, 180-191, 1999

[25]   P. Seitz, T. Spirig, "Vorrichtung und Verfahren zur Detektion und Demodulation eines intensitätsmodulierten Strahlungsfeldes," German Patent DE 4440613C1, Application 14.11.1994, Publication 25.07.1996

[26]   SR4000 Datasheet, MESA Imaging AG, Zürich, Switzerland, 2011

[27]   CamCube 3.0 Datasheet, PMD Technologies, Siegen, 2010

[28]   T. Ringbeck, "A 3D time-of-flight camera for object detection," Optical 3-D Measurement Techniques, ETH Zürich, 2007

[29]   R. Schwarte, "Verfahren und Vorrichtung zur Bestimmung der Phasen- und/oder Amplitudeninformation einer elektromagnetischen Welle," German Patent DE19704496, 1998.

[30]   R. Lange, "3D time-of-flight distance measurement with custom solid-state image sensors in CMOS/CCD-technology," PhD Thesis, Universität Siegen, Germany, 2000

[31]   B. Buttgen, Conference Discussion, International Workshop on Range-Imaging Sensors and Applications (RISA 2011), Trento, Italy, 2011

[32]   B. Buttgen, P. Seitz, "Robust optical time-of-flight range imaging based on smart pixel structures," IEEE Trans. Circuits and Systems, 55, 1512-1525, 2008

[33]   J.A. Richardson, L.A. Grant, R. K. Henderson, "Low dark count single-photon avalanche diode structure compatible with standard nanometer scale CMOS technology," IEEE Photonics Technology Lett., 21, 1020-1022, 2009

[34] D.-U. Li, J. Arlt, J. Richardson, R. Walker, A. Buts, D. Stoppa, E. Charbon, R. Henderson, "Real-time fluorescence lifetime imaging system with a 32 × 32 0.13 µm CMOS low dark-count single-photon avalanche diode array," Opt. Exp., 18, 10257-10269, 2010

[35] J.A. Richardson, "Time resolved single photon imaging in nanometer scale CMOS technology," PhD Thesis, University of Edinburgh, United Kingdom, 2010

[36] WaveMaster 8 Zi-A Datasheet, LeCroy Corporation, United States, 2011

[37] M. Cohen, F. Roy, D. Herault, Y. Cazaux, A. Gandolfi, J. P. Reynard, C. Cowache, E. Bruno, T. Girault, J. Vaillant, F. Barbier, Y. Sanchez, N. Hotellier, O. LeBorgne, C. Augier, A. Inard, T. Jagueneau, C. Zinck, J. Michailos, E. Mazaleyrat, "Fully optimized Cu based process with dedicated cavity etch for 1.75µm and 1.45µm pixel pitch CMOS image sensors," International Electron Devices Meeting (IEDM '06), 1-4, 2006

[38] C. Niclass, C. Favi, T. Kluter, F. Monnier, E. Charbon, "Single-photon synchronous detection," 34th European Solid-State Circuits Conference (ESSCIRC 2008), 114-117, 2008

[39] C. Niclass, C. Favi, T. Kluter, F. Monnier, E. Charbon, "Single-photon synchronous detection," IEEE J. Solid-State Circuits, 44, 1977-1989, 2009

[40] D. Piatti, "SR-4000 and CamCube3.0 ToF cameras: test and comparison," International Workshop on Range-Imaging Sensors and Applications (RISA 2011), Trento, Italy, 2011

[41] A. Rochas, "Single photon avalanche diodes in CMOS technology," PhD Thesis no. 2814, École Polytechnique Fédérale de Lausanne, Switzerland, 2003

[42] G. Grau, W. Freude, "Optische Nachrichtentechnik," 3. Auflage, Springer, Berlin, 1991

[43] M. A. Green, M. J. Keevers, "Optical properties of intrinsic silicon at 300 K," Progress in Photovoltaics: Research and Applications, 3, 189-192, 1995

[44] S. Cova, M. Ghioni, A. Lacaita, C. Samori, F. Zappa, "Avalanche photodiodes and quenching circuits for single-photon detection," Appl. Opt., 35, 1956-1976, 1996

[45] S. Pellegrini, R. E. Warburton, L. J. J. Tan, N. Jo Shien, A. B. Krysa, K. Groom, J. P. R. David, S. Cova, M. J. Robertson, G. S. Buller, "Design and performance of an InGaAs-InP single-photon avalanche diode detector," IEEE J. Quantum Electronics, 42, 397-403, 2006

[46] A.Y. Loudon, P.A. Hiskett, G.S. Buller, R.T. Carline, D.C. Herbert, W.Y. Leong, J.G. Rarity, "Enhancement of the infrared detection efficiency of silicon photon-counting avalanche photodiodes by use of silicon germanium absorbing layers," Opt. Lett., 27, 219-221, 2002

[47] E. Charbon, P. Seitz, "Non-conventional imaging," Solid State Imaging: Architectures and Techniques, EPFL Lecture, 2008

[48]   B.F. Aull, A.H. Loomis, D.J. Young, R.M. Heinrichs, B.J. Felton, P.J. Daniels, D.J. Landers, "Geiger-mode avalanche photodiodes for three-dimensional imaging," Lincoln Laboratory Journal, 12, 335-350, 2002

[49]   P. Seitz, "Quantum-noise limited distance resolution of optical range imaging techniques," IEEE Trans. Circuits and Systems, 55, 2368-2377, 2008

[50]   B. Saleh, M.C. Teich, "Fundamentals of Photonics," John Wiley and Sons, New York, 1991

[51]   D. Meschede, "Optik, Licht und Laser," Teubner, Stuttgart, 1999

[52]   M. Fox, "Quantum photonics - an introduction", Oxford University Press, New York, 2006

[53]   A. Eisele, R. Henderson, B. Schmidtke, T. Funk, L. Grant, J. Richardson, W. Freude, "185 MHz count rate, 139 dB dynamic range single-photon avalanche diode with active quenching circuit in 130 nm CMOS technology," International Image Sensor Workshop (IISW 2011), Hokkaido, Japan, 2011

[54]   R.H. Haitz, "Mechanisms contributing to the noise pulse rate of avalanche diodes," J. of Appl. Phys., 36, 3123-3131, 1965

[55]   W. Shockley, W.T. Read, "Statistics of the recombinations of holes and electrons," Physical Review, 87, 835, 1952

[56]   R. Henderson, J. Richardson, L. Grant, "Reduction of band-to-band tunneling in deep-submicron CMOS single-photon avalanche photodiodes," International Image Sensor Workshop (IISW 2009), Bergen, Norway, 2009

[57]   L. Pancheri, D. Stoppa, "Low-noise CMOS single-photon avalanche diodes with 32 ns dead time," 37th European Solid State Device Research Conference (ESSDERC 2007), 362-365, 2007

[58]   C. Niclass, M. Soga, "A miniature actively recharged single-photon detector free of afterpulsing effects with 6ns dead time in a 0.18µm CMOS technology," IEEE Int. Electron Devices Meeting (IEDM 2010), 14.3.1-14.3.4., 2010

[59]   S. Cova, A. Lacaita, G. Ripamonti, "Trapping phenomena in avalanche photodiodes on nanosecond scale," IEEE Electron. Device Lett., 12, 685-687, 1991

[60]   M. Ghioni, S. Cova, A. Lacaita, G. Ripamonti, "New silicon epitaxial avalanche diode for single-photon timing at room temperature," Electronics Lett., 24, 1476-1477, 1988

[61]   E. Charbon, M. Fishburn, "Monolithic single photon avalanche diodes: SPADs," Single-Photon Imaging, p. 123-157, Springer, Berlin, 2011

[62]   A. Lacaita, S. Cova, M. Ghioni, F. Zappa, "Single-photon avalanche diode with ultrafast pulse response free from slow tails," IEEE Electron. Device Lett., 14, 360-362, 1993

[63]   H. Dautet, P. Deschamps, B. Dion, A.D. MacGregor, D. MacSween, R.J. McIntyre, C. Trottier, P.P. Webb, "Photon counting techniques with silicon avalanche photodiodes," Appl. Opt., 32, 3894-3900, 1993

[64]   J.A. Richardson, E.A.G. Webster, L.A. Grant, R.K. Henderson, "Scaleable single-photon avalanche diode structures in nanometer CMOS technology," IEEE Trans. Electron Devices, 58, 2028-2035, 2011

[65]   S. Donati, G. Martini, M. Norgia, "Microconcentrators to recover fill-factor inimage photodetectors with pixel on-boardprocessing circuits," Opt. Express, vol. 15, pp. 18066-18075, 2007.

[66]   A.L. Lacaita, F. Zappa, S. Bigliardi, M. Manfredi, "On the bremsstrahlung origin of hot-carrier-induced photons in silicon devices," IEEE Trans. Electron Devices, 40, 577-582, 1993

[67]   R.J. McIntyre, "Theory of microplasma instability in silicon," J. of Appl. Phys., 32, 983-995, 1961

[68]   R.H. Haitz, "Model for the electrical behavior of a microplasma," J. Appl. Phys., 35, 1370-1376, 1964

[69]   D. Renker, "Geiger-mode avalanche photodiodes, history, properties and problems," Nuclear Instruments and Methods in Physics Research Section A: Accelerators, Spectrometers, Detectors and Associated Equipment, 567, 48-56, 2006

[70]   G.A. Petrillo, R.J. McIntyre, R. Lecomte, G. Lamoureux, D. Schmitt, "Scintillation detection with large-area reach-through avalanche photodiodes," IEEE Trans. Nuclear Science, 31, 417-423, 1984

[71]   R.J. McIntyre, "Recent developments in silicon avalanche photodiodes," Measurement, 3, 146-152, 1985

[72]   Single Photon Counting Module, SPCM-AQR, PerkinElmer Optoelectronics, Fremont (CA), United States, 2004

[73]   S. Cova, A. Longoni, A. Andreoni, "Towards picosecond resolution with single-photon avalanche diodes," Review of Scientific Instr., 52, 408-412, 1981

[74]   S. Cova, A. Lacaita, M. Ghioni, G. Ripamonti, T.A. Louis, "20-ps timing resolution with single-photon avalanche diodes," Review of Scientific Instr., 60, 1104-1110, 1989

[75]   A. Lacaita, M. Ghioni, S. Cova, "Double epitaxy improves single-photon avalanche diode performance," Electronics Lett., 25, 841-843, 1989

[76]   A. Rochas, M. Gani, B. Furrer, P.A. Besse, R.S. Popovic, G. Ribordy, N. Gisin, "Single photon detector fabricated in a complementary metal-oxide-semiconductor high-voltage technology," Review of Scientific Instr., 74, 3263-3270, 2003

[77]   C. Niclass, M. Sergio, E. Charbon, "A Single photon avalanche diode array Fabricated in deep-submicron CMOS technology," Proc. Design, Automation and Test in Europe (DATE '06), pp. 1-6, 2006

[78]   C. Niclass, M. Sergio, E. Charbon, "A single photon avalanche diode array fabricated in 0.35µm CMOS and based on an event-driven readout for TCSPC experiments," SPIE Proc. Advanced Photon Counting Techniques, 6372, 2006

[79]   M.A. Marwick, A.G. Andreou, "Single photon avalanche photodetector with integrated quenching fabricated in TSMC 0.18 um 1.8 V CMOS process," Electronics Lett., 44, 643-644, 2008

[80]   C. Niclass, M. Gersbach, R. Henderson, L. Grant, E. Charbon, "A 130-nm CMOS single-photon avalanche diode," Proc. SPIE, 676606-9, Boston, United States, 2007

[81]   M.A. Karami, M. Gersbach, H.-J. Yoon, E. Charbon, "A new single-photon avalanche diode in 90nm standard CMOS technology," Opt. Express, 18, 22158-22166, 2010

[82]   C. Niclass, M. Gersbach, R. Henderson, L. Grant, E. Charbon, "A single photon avalanche diode implemented in 130-nm CMOS technology," IEEE J. Selected Topics in Quantum Electronics, 13, 863-869, 2007

[83]   E. Webster, J. Richardson, L. Grant, R. Henderson, "Single-photon avalanche diodes in 90nm CMOS imaging technology with sub-1Hz median dark count rate," IEEE International Image Sensor Workshop (IISW 2011), Hokkaido, Japan, 2011

[84]   E. Webster, J. Richardson, L. Grant, R. Henderson, "An infra-red sensitive, low noise, single-photon avalanche diode in 90nm CMOS," International Image Sensor Workshop (IISW 2011), Hokkaido, Japan, 2011

[85]   H. Finkelstein, M.J. Hsu, S.C. Esener, "STI-bounded single-photon avalanche diode in a deep-submicrometer CMOS technology," IEEE Electron. Device Lett., 27, 887-889, 2006

[86]   J. Richardson, R. Henderson, D. Renshaw, "Dynamic quenching for single photon avalanche diode arrays," IEEE International Image Sensor Workshop (IISW 2007), Ogunquit, United States, 2007

[87]   W.R. Leo, "Techniques for nuclear and particle physics experiments", Springer, Berlin, 1994

[88]   S.H. Lee, R.P. Gardner, "A new G-M counter dead time model," Applied Radiation and Isotopes, 53, 731-737, 2000

[89]   L. Neri, S. Tudisco, F. Musumeci, A. Scordino, G. Fallica, M. Mazzillo, M. Zimbone, "Note: dead time causes and correction method for single photon avalanche diode devices," Review of Scientific Instr., 81, 086102-086102-3, 2010

[90]   T. Murphy, "Statistics of photon arrival time," APOLLO Project Documents, 2001.

[91]   U. Kienke, H. Jäkel, "Signale und Systeme," Oldenburg, München, Germany, 2005

[92]   W. Becker, H. Hickl, "The bh TCSPC Handbook," Berlin, Germany, 2006

[93]   Wikipedia, "Superellipse," http://en.wikipedia.org/wiki/Superellipse, 13.09.2011

[94]   D. Mosconi, D. Stoppa, L. Pancheri, L. Gonzo, A. Simoni, "CMOS single-photon avalanche diode array for time-resolved fluorescence detection," 32nd European Solid-State Circuits Conference (ESSCIRC 2006), 564-567, 2006

[95]   R.K. Henderson, E.A.G. Webster, R. Walker, J.A. Richardson, L.A. Grant, "A 3x3, 5µm pitch, 3-transistor single photon avalanche diode array with integrated 11V bias generation in 90nm CMOS technology," IEEE Int. Electron Devices Meeting (IEDM 2010), 14.2.1-14.2.4., 2010

[96]   A. Eisele, B. Schmidtke, R. Schnitzer, „Entfernungsmessgerät mit homogenisierender Messauswertung," German patent application DE102010003843A1, 2011

[97]   A. Eisele, R. Schnitzer, B. Schmidtke, P. Wolf, "Entfernungsmessgerät mit Homogenisierender Messauswertung," German patent application DE102010039295A1, 2012

[98]   L. Möller, "Akzeptanz von Solaranlagen: Institut für Geographie und Geoökologie," Universität Karlsruhe, 1999

[99]   G. Kramm, N. Mölders, "Planck's blackbody radiation law: presentation in different domains and determination of the related dimensional constants," J. Calcutta Mathematical Society, 2009

[100] Laser Line Filter 10LF01-633 Datasheet, Newport Corporation, Irvine (CA), United States, 2011

[101] Wikipedia, "Lambertian Diffuse Lighting Model", http://en.wikipedia.org/wiki/Lambertian_diffuse_lighting_model, 06.10.2011

[102] K. Iga, "Surface-emitting laser-its birth and generation of new optoelectronics field," IEEE J. Selected Topics in Quantum Electronics, 6, 1201-1215, 2000

[103] B. Schmidtke, "Halbleiterlaserdynamik," Bosch Presentation, 2010

[104] Leica Disto D2 Device and Manual, Leica Geosystems AG, Heerbrugg, Switzerland, 2008

[105] Leica Disto D8 Device and Manual, Leica Geosystems AG, Heerbrugg, Switzerland, 2011

[106] ADL-63054TL Laser Diode Datasheet, Arima Laser, Taiwan, 2005

[107] A. Eisele, "Burst Modulation - Laser Safety," Bosch Presentation, 2010

[108] P. Forrer, K. Giger, "Optical-electronic distance measuring device," Patent Application WO 2009/101002 A1, Application 11.02.2008, Publication 20.08.2009

[109] R. Martinsen, K. Kennedy, A. Radl, "Speckle in laser imagery: efficient methods of quantification and minimization," 12th Annual Meeting IEEE Lasers and Electro-Optics Society (LEOS '99), 354-355, 1999

[110] E. Hecht, "Optics", 4 ed., Addison Wesley, 2002

[111] M. Richard, C. Pahud, B. Haase, "Optical receiver lens and optical distance measuring device," PCT Patent Application WO002010069633A1, Application 17.12.2008, Publication 24.06.2010

[112] B. Schmidtke, "Raytracing simulation of laser rangefinder with zemax," Bosch Simulation, 2011

[113] T.M. Souders, D.R. Flach, C. Hagwood, G.L. Yang, "The effects of timing jitter in sampling systems," IEEE Trans. Instr. and Measurement, 39, 80-85, 1990

[114] G. Metzler, "Fehlerkorrigiertes 50-GHz-Messsystem für nichtlineare elektrooptische Ein- und Zweitore," PhD Thesis, Universität Karlsruhe, Germany, 2000

[115] M. Hsieh, G.E. Sobelman, "Comparison of LC and ring VCOs for PLLs in a 90 nm digital CMOS process," Proc. Int. SoC Design Conference, 19-22, 2006

[116] R. Behzad, "Design of integrated circuits for optical communications", McGraw-Hill Science, 2002

[117] T. Miyazaki, M. Hashimoto, H. Onodera, "A performance comparison of PLLs for clock generation using ring oscillator VCO and LC oscillator in a digital CMOS process," Proc. ASP-DAC, 545-546, 2004

[118] L. Fanucci, R. Roncella, R. Saletti, "Non-linearity reduction technique for delay-locked delay-lines," IEEE Int. Symposium on Circuits and Systems (ISCAS 2001), 4, 430-433, 2001

[119] C.-K.K. Yang, "Delay-locked loops - an overview," Phase-Locking in High-Performance Systems, B. Razavi, p. 13-22, Wiley-IEEE Press, 2003

[120] F. Baronti, L. Fanucci, D. Lunardini, R. Roncella, R. Saletti, "On-line calibration for non-linearity reduction of delay-locked delay-lines," 8th IEEE Int. Conf. on Electronics, Circuits and Systems (ICECS 2001), 2, 1001-1005, 2001

[121] M.J.M. Pelgrom, A.C.J. Duinmaijer, A.P.G. Welbers, "Matching properties of MOS transistors," IEEE J. Solid-State Circuits, 24, 1433-1439, 1989

[122] F. Lustenberger, T. Oggier, G. Becker, L. Lamesch, "Method and device for redundant distance measurement and mismatch cancelation in phase-measurement systems," EP1752793, 2005

[123] R. Lange, P. Seitz, A. Biber, S. C. Lauxtermann, "Demodulation pixels in CCD and CMOS technologies for time-of-flight ranging," Proc. SPIE 3965, Sensors and Camera Systems for Scientific, Industrial, and Digital Photography Applications, San Jose (CA), United States, 177-188, 2000

[124] J. Doernberg, L. Hae-Seung, D. Hodges, "Full-speed testing of A/D converters," IEEE J. Solid-State Circuits, 19, 820-827, 1984

[125] F. Baronti, D. Lunardini, R. Roncella, R. Saletti, "Experimental characterization of a self-calibrating delay-locked delay-line," Southwest Symposium on Mixed-Signal Design, pp. 94-98, 2003

[126] M. Mota, J. Christiansen, "A high-resolution time interpolator based on a delay locked loop and an RC delay line," IEEE J. of Solid-State Circuits, 34, 1360-1366, 1999

[127] B. Schmidtke, O. Wolst, A. Eisele, "Optisches Entfernungsmessgerät mit Kalibriereinrichtung," German patent application DE102009045323A1, 2011

[128] A. Eisele, B. Schmidtke, "Optisches Entfernungsmessgerät mit Kalibriereinrichtung zum Berücksichtigen von Übersprechen," German patent application DE, Application 2012

[129] Technical Documents and Simulation Models for IMG175 130nm CMOS Technology, ST Microelectronics, Edinburgh, United Kingdom, 2011

[130] On-Chip High-Voltage Generation Concept, STMicroelectronics, 2010

[131] Bronstein, Semendjajew, Musiol, Mühlig, "Taschenbuch der Mathematik," 5. ed., p. 476, Harri Deutsch, 2001

Acknowledgements

This thesis is the result of my work at the Institute of Photonics and Quantum Electronics (IPQ) at the Karlsruhe Institute of Technology and at Robert Bosch GmbH (Power Tools / Measuring Tools / PT-MT/ENG) from October 2008 to November 2011.

I would like to express my gratitude particularly to my academic supervisor at the Institute of Photonics and Quantum Electronics (IPQ) at the Karlsruhe Institute of Technology, Prof. Wolfgang Freude, and my industrial supervisor at Robert Bosch GmbH, Dr. Bernd Schmidtke, who guided this research.

My academic supervisor, Prof. Wolfgang Freude, I would like to thank for his analytic expertise and his scientific thoroughness, which complemented the more application-oriented view at Bosch. Despite his busy schedule he was always ready for helpful and clarifying discussions. Without his engagement this work would not have been accomplished in its present form.

I would also like to express my thanks to Prof. Jürg Leuthold, Head of IPQ until February 2013 (now with Swiss Federal Institute of Technology, Zürich), for his encouragement and for kindly providing local resources for my work.

My industrial supervisor, Dr. Bernd Schmidtke, a brilliant physicist, I would like to thank for his patient guidance, enthusiastic encouragement and useful criticism of this research work. I am particularly grateful for his openess towards new ideas, and for his willingness to give them a try in the laboratory: The working environment at Bosch could not have been better.

I would like to thank Wolfgang Baierl, head of the engineering division of Bosch Measuring Tools as well as Dr. Oliver Wolst and Dr. Ulrich Kallmann as my group leaders at Bosch. Without their entrepreneurial spirit and willingness to take a risk for a promising project, this work would not have been possible. In particular during the early stages Oliver's strong support was the foundation of this project.

I would like to express my very great appreciation to Reiner Schnitzer from Bosch Automotive Electronics, whose sound knowledge of semiconductor fabrication and ASIC design helped theoretical ideas come to life. He has been with this project since the early days and will certainly ensure that the concepts of this thesis will find their way into a successful commercial product.

I would also like to thank my dear colleagues Florian Giesen and Steffen Assmann as well as the entire laser rangefinder group from Bosch Measuring Tools who would always be open for questions and more than once provided a helping hand in electronics development and programming. Tobias Funk greatly contributed to an almost completely automated laboratory setup using Matlab. The assistance provided at the machine shop for experimental setups is highly appreciated. My special thanks are extended to Stephan Hetterscheidt from the Intellectual Property department at Bosch. Not only did he ensure the patent protection of the core ideas but also provided guidance for my future professional career.

For the great collaboration on SPADs, I want to thank the people at STMicroelectronics (Imaging Division, UK, Edinburgh) as well as Robert Henderson and Justin Richardson from the University of Edinburgh. Their work on single-photon avalanche diodes (SPADs) is the technical foundation for counting billions of photons throughout this research. The warm welcome in Edinburgh and the open atmosphere during technical discussions are very much appreciated.

Many thanks also to all further colleagues at Bosch, in particular for the most valuable 'coffee discussions' with Martin Pohlmann, Jan Brosi, Heiko Braun and Dorothea Sturtz.

Last but not least I wish to thank my family for their constant support and encouragement.

List of Own Publications and Patent Applications

1. A. Eisele, R. Henderson, B. Schmidtke, T. Funk, L. Grant, J. Richardson, W. Freude, "185 MHz count rate, 139 dB dynamic range single-photon avalanche diode with active quenching circuit in 130 nm CMOS technology," International Image Sensor Workshop (IISW 2011), Hokkaido, Japan, 2011

2. A. Eisele, "Photon-counting laser rangefinder," seminar at the Center for Research and Education in Optics and Lasers (CREOL), UCF, USA, April 2011

3. B. Schmidtke, O. Wolst, A. Eisele, "Messvorrichtung zur Messung einer Entfernung zwischen der Messvorrichtung und einem Zielobjekt mit Hilfe optischer Messstrahlung," German patent application DE102009029372A1, 2011

   B. Schmidtke, O. Wolst, A. Eisele, "Optical Measuring Device," PCT patent application WO002011029645A1, 2011

   Patent pending in the US as US20120249998, in Europe as EP2475958A1, and in China as CN102549381A

4. B. Schmidtke, O. Wolst, A. Eisele, "Messvorrichtung zur Messung einer Entfernung zwischen der Messvorrichtung und einem Zielobjekt mit Hilfe optischer Messstrahlung," German patent application DE102009029364A1, 2011

   B. Schmidtke, O. Wolst, A. Eisele, "Optical Measuring Device, " PCT patent application WO002011029651A1, 2011

   Patent pending in the US as US20120262696, in Europe as EP2475957A1, and in China as CN102549380A

5. A. Eisele, B. Schmidtke, O. Wolst, "Photonendetektor mir paralysierbarem Photonenempfindlichen Element, insbesondere SPAD, sowie Entfernungsmessgerät mit solchem Photonendetektor," German patent application DE102009029376A1, 2011

   A. Eisele, B. Schmidtke, O. Wolst, "Photon detector with an immobilisable photonsensitive element, in particular SPAD and distancing measuring device comprising said type of photon detector," PCT patent application WO002011029646A1, 2011

   Patent pending in the US as US20120261547, in Europe as EP2476013A1, and in China as CN102576071A

6. B. Schmidtke, O. Wolst, A. Eisele, "Optisches Entfernungsmessgerät mit Kalibriereinrichtung," German patent application DE102009045323A1, 2011

   B. Schmidtke, O. Wolst, A. Eisele, "Optical distance measuring device with calibration device," PCT patent application WO002011042290A1, 2011

   Patent pending in the US as US20120236290, in Europe as EP2486370A1, and in China as CN102656423A

7. A. Eisele, B. Schmidtke, R. Schnitzer, "Entfernungsmessgerät mit homogenisierender Messauswertung," German patent application DE102010003843A1, 2011

A. Eisele, B. Schmidtke, R. Schnitzer, "Distance measuring device having a homogenizing measurement evaluation," PCT patent application WO002011128131A1, 2011

Patent pending in the US as US20130208258, in Europe as EP2558883A1, and in China as CN102844676A

8. A. Eisele, R. Schnitzer, B. Schmidtke, P. Wolf, "Entfernungsmessgerät mit Homogenisierender Messauswertung," German patent application DE102010039295A1, 2012

9. A. Eisele, O. Wolst, B. Schmidtke, U. Skultety-Betz, "Optische Abschirm-Vorrichtung zum Trennen von optischen Pfaden," German patent application DE102010041937A1, 2012

A. Eisele, O. Wolst, B. Schmidtke, U. Skultety-Betz, "Optical shielding device for separating optical paths," PCT patent application WO002012045503A1, 2012

Patent pending in Europe as EP2625545A1, and in China as CN103124912A

10. A. Eisele, B. Schmidtke, R. Schnitzer, "Messvorrichtung zur Messung einer Entfernung zwischen der Messvorrichtung und einem Zielobjekt mit Hilfe optischer Messstrahlung," German patent application DE102011005740A1, 2012

A. Eisele, B. Schmidtke, R. Schnitzer, "Measurement device for measuring a distance between the measurement device and a target object using an optical measurement beam," PCT patent application WO2012123152A1, 2012

11. B. Schmidtke, A. Eisele, U. Kallmann, "Messvorrichtung zur Mehrdimensionalen Vermessung eines Zielobjekts," German patent application DE102011005746A1, 2012

B. Schmidtke, A. Eisele, U. Kallmann, "Measuring apparatus and measuring device for measuring a target object in a multidimensional manner," WO2012126659A1, 2012

12. A. Eisele, "Entfernungsmessvorrichtung," German patent application DE102011089328A1, 2013

A. Eisele, "Distance measuring device and method for distance measurement," PCT patent application WO2013091984A1, 2013

13. A. Eisele, "Entfernungsmessvorrichtung," German patent application DE102011089824A1, 2013

A. Eisele, "Distance measuring device and method for distance measurement," PCT patent application WO2013091985A1, 2013

14. A. Eisele, "Verfahren zur Abtastung eines sich bewegenden Fahrzeugs," German patent application DE102012200336A1, 2013

    A. Eisele, "Method for scanning the environment of a moving vehicle," PCT patent application WO2013104459A1, 2013

15. A. Eisele, "Verfahren zur Steuerung mindestens eines Aktors anhand von Signallaufzeitänderungen mindestens eines Ultraschallsensors," German patent application DE102011087347A1, 2013

    A. Eisele, "Process for controlling at least one actuator on the basis of changes of signal transit-time of at least one ultrasonic sensor," UK patent application GB2498833A, 2013

    A. Eisele, "Procédé de commande d'un actionneur à partir des variations du temps de parcours de signal fourni par un capteur à ultrasons," French patent application FR2983308A1, 2013

Further patent applications are about to be published.

Curriculum Vitae

Andreas Eisele

born March 26, 1985

in Stuttgart, Germany

Citizenship: German

Education

2008-2013	PhD Project "Millimeter-Precision Laser Rangefinder using a Low-Cost Photon Counter", Institute of Photonics and Quantum Electronics (IPQ), Karlsruhe Institute of Technology, Germany
2007-2008	M.Sc. Optics, Center for Research and Education in Optics and Lasers (CREOL), University of Central Florida, USA
2004-2007	Studies of Electrical Engineering and Information Technology, Karlsruhe Institute of Technology, Germany

Internships and Work Experience

10/2008-11/2011	Research Scientist, Engineering Measuring Tools (PT-MT/ENG), Robert Bosch GmbH
08/2007-09/2008	Graduate Research Assistant, Optical Fiber Communications Group, Center for Research and Education in Optics and Lasers (CREOL), University of Central Florida, USA
11/2004-08/2007	Consulting and IT administration with own firm Power Consult GbR
04/2007-07/2007	Teaching Assistant, Light Technology Institute (LTI), Karlsruhe Institute of Technology
03/2006-07/2006	Teaching Assistant, Institute of Biomedical Engineering (IBT), Karlsruhe Institute of Technology
10/2005-11/2005	Internship at Porsche AG
03/2005-05/2005	Internship at Robert Bosch GmbH

Scholarships

Fulbright Commission, Studienstiftung des Deutschen Volkes, Deutsche Telekom Stiftung